Fundamentals of Sustainable Aviation Fuels

Fundamentals of Sustainable Aviation Fuels

Nadir Yilmaz, Ph.D., P.E.

400 Commonwealth Drive
Warrendale, PA 15096-0001 USA
E-mail: CustomerService@sae.org
Phone: 877-606-7323 (inside USA and Canada)
724-776-4970 (outside USA)
Fax: 724-776-0790

Library of Congress Catalog Number 2025943016
http://dx.doi.org/10.4271/9781468607925

ISBN-Print 978-1-4686-0791-8
ISBN-PDF 978-1-4686-0792-5
ISBN-epub 978-1-4686-0793-2

To purchase bulk quantities, please contact: SAE Customer Service

E-mail: CustomerService@sae.org
Phone: 877-606-7323 (inside USA and Canada)
724-776-4970 (outside USA)
Fax: 724-776-0790

Visit the SAE International Bookstore at books.sae.org

Publisher
Sherry Dickinson Nigam

Product Manager
Amanda Zeidan

Production and Manufacturing Associate
Michelle Silberman

To my son, Evren Oisin, who is my greatest inspiration

Contents

Introduction

Energy impacts every aspect of industrial and social life, highlighting the need to alter production and consumption practices, improve the related resource use, and implement effective resource management across all sectors. Sustainable and efficient management of resources can be achieved by using fewer feedstocks, increasing resource efficiency, reducing waste generation, and recycling waste into resources, i.e., by promoting sustainability. Building a sustainable future requires implementing significant energy initiatives and taking concrete steps toward progress. The transportation sector is at the forefront of these necessary steps. The sustainability momentum initiated in road transportation has also become a necessity for air transportation. The aviation industry is completely dependent on fossil fuels. As the demand for passenger transportation increases, so does the consumption of liquid fuels. One of the sustainability issues with the highest priority in the aviation sector is alternative fuels. Sustainability in the aviation sector is implemented through initiatives encouraging alternatives to fossil fuels, the primary source of energy consumption—mainly by environmental regulations targeting airline companies. Developing an alternative aviation fuel suitable for global use depends on the following interconnected factors: sustainable feedstock supplies, effective and affordable production technology, safety and suitability testing, rapid transportation through existing pipelines or supply lines, and fuel blend creation. It has become inevitable for international control, regulation, and monitoring mechanisms to emerge in order to manage and promote the production of fuels that will serve as alternatives to aviation fuels. At this point, basic administrative organizations such as The International Civil Aviation Organization (ICAO) and International Air Transport Association (IATA) have taken the lead in the initiative for sustainable aviation fuels (SAFs). These organizations, which have a say in both environmental and technological developments of the aviation sector as policymakers, have followed a strict policy worldwide and ensured that countries have their own sustainable aviation roadmaps. The US, via its Sustainable Aviation Fuel Grand Challenge, is the leader in air transportation in terms of population density and the number of flights, the European Union (ReFuelEU Aviation) policies are the pioneers of environmental norms, the regulations are determined by the UK (SAF Mandate) and Canada (C-SAF Roadmap), while South American and Asian countries are trying to determine future projections by expanding the mandatory use of SAFs.

Today, advancing the production and widespread airport use of SAFs through academic and industrial collaboration requires understanding past experiences, clearly defining present conditions, and documenting emerging technological developments. Currently, academic research on SAFs is limited. While multi-authored books cover various components of the

SAF production chain, these publications repeatedly follow similar structures, beginning from identical starting points and concluding with broad, generalized results. The number of books in the literature that deal with the processes involved in the production of SAFs along with the projections for the future is also limited. Therefore, it is necessary to further investigate the production of alternative fuels made of oil and waste, their expansion, and their utilization in the aviation industry. The civil aviation sector actively implements all regulations regarding SAF use, and countries are forming joint agreements to ensure consistent SAF quality at every destination. In military aviation, there has been no obligation to use alternative fuels until recently. However, NATO's inclusion of SAFs in its single-fuel regulation for member countries will facilitate the use and further development of SAF technology across all stages of the SAF supply chain. In this respect, the information provided in this book on SAF usage in civil aviation serves as a foundational resource for similar studies in military aviation. Thus, to accurately project the future fuel requirements of this sector, presenting comprehensive literature on the topic is essential.

With this motivation in mind, this book was prepared to support the industry's efforts toward using SAF and explain sustainable aviation concepts to a wider audience, which is presented to the readers in five chapters covering the following: the importance of SAFs and current policies **(Chapter 1)**; production pathways to SAFs including feedstocks, conversion and production process, and production technologies **(Chapter 2)**; environmental impact of SAFs, combustion performance, emissions, and the potential of reduction of greenhouse gases (GHGs) **(Chapter 3)**; usage of SAFs, drop-in fuels in terms of safety and cost, demand and availability, specific fuel requirements, and fuel systems **(Chapter 4)**; and future of SAFs, progress and regulation status, and forthcoming activities **(Chapter 5)**.

Chapter 1 highlights the need to use new and renewable energy sources on a sectoral basis, based on energy production worldwide and dependency rates according to the area of use of resources. The aviation sector, which increasingly influences transportation, will require more liquid fuel in the near future. Chapter 1 explains in detail the importance and current status of using SAFs to meet this need, particularly in terms of complying with strict environmental regulations and reducing dependency on fossil fuels. Current steps taken by developed countries to shift their dependence on fossil fuels to fuels that have the same properties as fossil fuels and can be produced from sustainable sources to protect the environment are also outlined.

Chapter 2 introduces the basic components of the regional classification of feedstocks suitable for SAF production and efficient fuel production methods suitable for the feedstock, and it also explains the infrastructure of production technologies that are becoming widespread within the existing technologies. The criteria of high production, inedibility, accessibility, and suitability for fuel production methods are fundamental to the sustainability of feedstocks. In this chapter, all types of feedstocks, from the first to the fourth generation, are examined in detail within the scope of these criteria. The selection of SAF production methods according to their transformation potential based on the content of the raw materials and their compliance with sustainability and the current Advancing Standards Transforming Market (ASTM) standards are summarized. Within the Carbon Offsetting and Reduction Scheme for International Aviation (CORSIA) framework, SAF-approved conversion processes and the resulting fuel's mixing ratios according to ASTM standards are also indicated. The production methods of SAFs suitable for high mixing ratios are highlighted, and a comparison is made between long-term feedstock production with a low-cost process and efficient production technologies.

In **Chapter 3**, the environmental impacts of SAFs are discussed, and the environmental effects of emissions from the aviation sector are presented, indicating the place of SAFs in current emission reduction policies. Performance parameters in the existing literature are presented according to the operating principles of gas turbine engines, and their effects are evaluated together with end-of-combustion pollutants. Additionally, the reduction in emissions from aircraft and climate impacts due to flight patterns and contrail formation are also addressed with the use of SAFs. The chapter also evaluates the potential GHG-saving effect of SAFs from a different perspective.

Chapter 4 discusses the drop-in use of SAFs by addressing ASTM standards regarding essential fuel properties for safe use in aircraft. The role of market-leading countries in expanding SAF production and drop-in use is emphasized. As the need for SAF increases as a natural result of the supply agreements made worldwide and the potential of industrial production companies, issues such as easy access to SAFs, transportation safety, and the ability to cover the total cost at the end of the mixture formation are detailed. It is emphasized chronologically that due to the widespread use of SAFs, control and quality policies in production are provided with a strict specific fuel requirement and that the recording of processes as a single certification section by ASTM is accepted worldwide.

In **Chapter 5**, the projection of SAFs on the world map is shown from a perspective that brings producers and consumers together. Considering the roadmaps and legal regulations regarding the use of SAFs in developed countries and the guidelines currently implemented by the leading aviation organizations, the place of SAFs among the targets for the near future is revealed. In line with the goals of companies aiming to produce SAFs worldwide, initiatives to ensure that the mixing ratios of SAFs meet the standards at the targeted times are also discussed.

To conclude, although it is a separate topic of discussion that hydrogen and electrification technologies cannot yet become widespread in aircraft, it has been emphasized that using SAFs will be a promising approach in the aviation sector in the short and long term. Thus, following a holistic approach, this book provides excellent up-to-date guidance as a technical handbook in terms of understanding the importance of SAFs, the classification of feedstocks that can be used in SAF production, production methods according to feedstocks, compliance with environmental legislation, suitability for current airport and aircraft engine technologies, elimination of environmental risk factors, and transition to drop-in use by gradually increasing the mixing ratios.

List of Acronyms

APAC - Asia Pacific

ASTM - American Society for Testing and Materials

ATAG - Air Transport Action Group

AtJ - Alcohol to Jet

AtJ-SPK - Alcohol to Jet Synthetic Paraffinic Kerosene

BE - Battery Electric

BP - British Petroleum

CAAFI - Commercial Aviation Alternative Fuels Initiative

CAEP - Committee on Aviation Environmental Protection

CAGR - Compound Annual Growth Rate

CAPEX - Capital Expenditures

CEF - CORSIA-Eligible Fuels

CFPC - Clean Fuels Production Credit

CH-SK/CHJ - Catalytic Hydrothermolysis-Synthesized Kerosene/Jet

CI - Carbon Intensity

CO - Carbon Monoxide

CO_2 - Carbon Dioxide

COA - Certificate of Analysis

CORSIA - Carbon Offsetting and Reduction Scheme for International Aviation

CtL - Coal to Liquid

DEF STAN - Defence Standard

DSHC - Direct Sugar to Hydrocarbon

EASA - European Union Aviation Safety Agency

e-Fuel - Electro Synthetic Fuel

EIA - Energy Information Administration

EPA - Environmental Protection Agency

ETS - Emissions Trading Scheme

EU - European Union

FAA - Federal Aviation Administration

FAME - Fatty Acid Methyl Ester

FAST - Fuels for Sustainable Aviation Transition

FC - Fuel Cell

FEL - Flight Emission Label

FOG - Fats, Oils, and Greases

FQD - Fuel Quality Directive

FRL - Fuel Readiness Level

FT - Fischer–Tropsch

FT-SPK - Fischer–Tropsch Synthetic Paraffinic Kerosene

FT-SPK/A - Fischer–Tropsch Synthetic Paraffinic Kerosene with Aromatics

GFAAF - Global Aviation Alternative Fuels Framework

GHG - Greenhouse Gas

GtL - Gas to Liquid

HC - Hydrocarbon

HEFA - Hydroprocessed Esters and Fatty Acid

HEFA-SPK - Hydroprocessed Esters and Fatty Acids to Synthetic Paraffinic Kerosene

HFS-SIP - Hydroprocessed Fermented Sugars to Synthetic Isoparaffin

HRJ - Hydroprocessed Renewable Jet

HVO - Hydrogenated Vegetable Oil

IATA - International Air Transport Association

ICAO - International Civil Aviation Organization

ICE - Internal Combustion Engine

IEA - International Energy Agency

ILUC - Indirect Land-Use Change

JP - Jet Propellant

LCA - Lifecycle assessment

LCFS - Low Carbon Fuel Standard

LTO - Landing–Takeoff

MENA - Middle East and North Africa

MOD - Ministry of Defence

MSW - Municipal Solid Waste

MtG - Methanol to Gasoline

NASA - National Aeronautics and Space Administration

NO_x - Nitrogen Oxide

OPEX - Operating Expenses

PAH - Polycyclic Aromatic Hydrocarbon

PEMFC - Proton Exchange Membrane Fuel Cell

PM - Particle Matter

PtL - Power to Liquid

RCQ - Refinery Certificate of Quality

RED - Renewable Energy Directive

RSB - Roundtable on Sustainable Biomaterials

RTK - Revenue Tonne Kilometers

SAF - Sustainable Aviation Fuel

SOFC - Solid Oxide Fuel Cell

SO_x - Sulfur Oxide

SPK - Synthetic Paraffinic Kerosene

StL - Sun to Liquid

TRL - Technology Readiness Level

UCO - Used Cooking Oil

UHC - Unburned Hydrocarbon

UK - United Kingdom

UNFCCC - United Nations Framework Convention on Climate Change

US - United States

VOC - Volatile Organic Compound

Chapter 01

SAFs

Energy is crucial for meeting human needs and driving economic development. As the world population is rapidly increasing, there is a greater need for primary energy sources day by day [1.1]. Global energy consumption is expected to increase gradually due to slowing economies and high energy prices [1.2]. Meeting these energy needs is essential for societal development. As the leading society in the level of development and industrialization, the United States (US) is the world's largest consumer of energy, in general, and of oil and refined products, in particular. The total energy consumption of the US, which ranks first in the developed countries category, is shown in **Figure 1.1**. The three major fossil fuels, petroleum (38%), natural gas (36%), and coal (9%), combined accounted for about 83% of primary energy production in the US in 2023 [1.3].

It is predicted that fossil sources will maintain their place as the main energy source worldwide in the near future, as it has in the past. The high use of fossil-based products clearly reveals the continued dependence on such energy sources. Regardless of the sector, the basic needs of everyday life require the highest energy consumption and continued supply. Although the production and manufacturing industry sectors have a large proportion of energy consumption, the transportation sector has the greatest need and proportion for constant energy supply and consumption, as depicted in **Figure 1.2** [1.4]. In energy projection studies conducted around the world, it is predicted that there will be no decrease in the sectoral distribution of primary energy consumption for industrial and transportation end-use sectors by 2050 [1.5].

Figure 1.1 US primary energy consumption by energy source, 2023 [1.3].

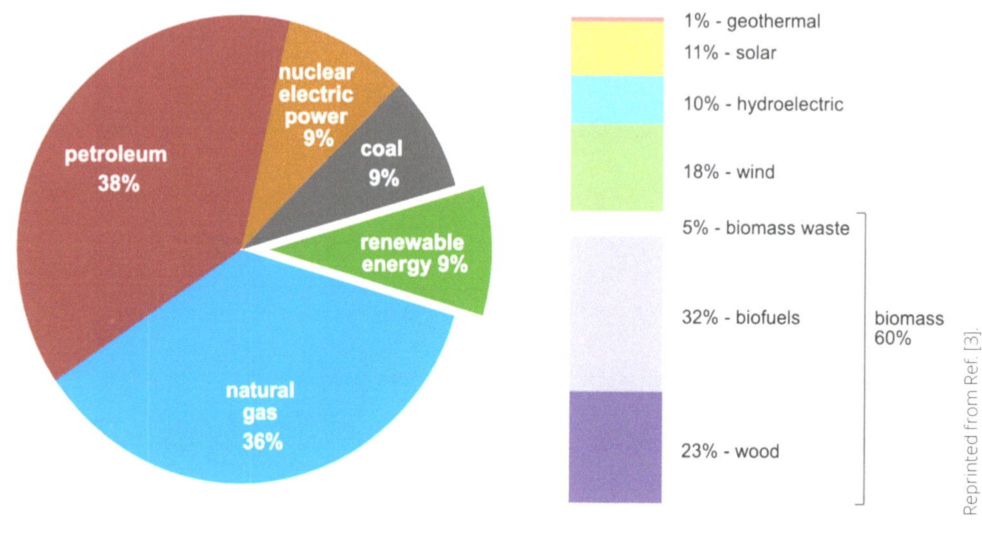

Reprinted from Ref. [3].

Figure 1.2 Primary energy consumption in the US by energy source and sector, 2023 [1.4].

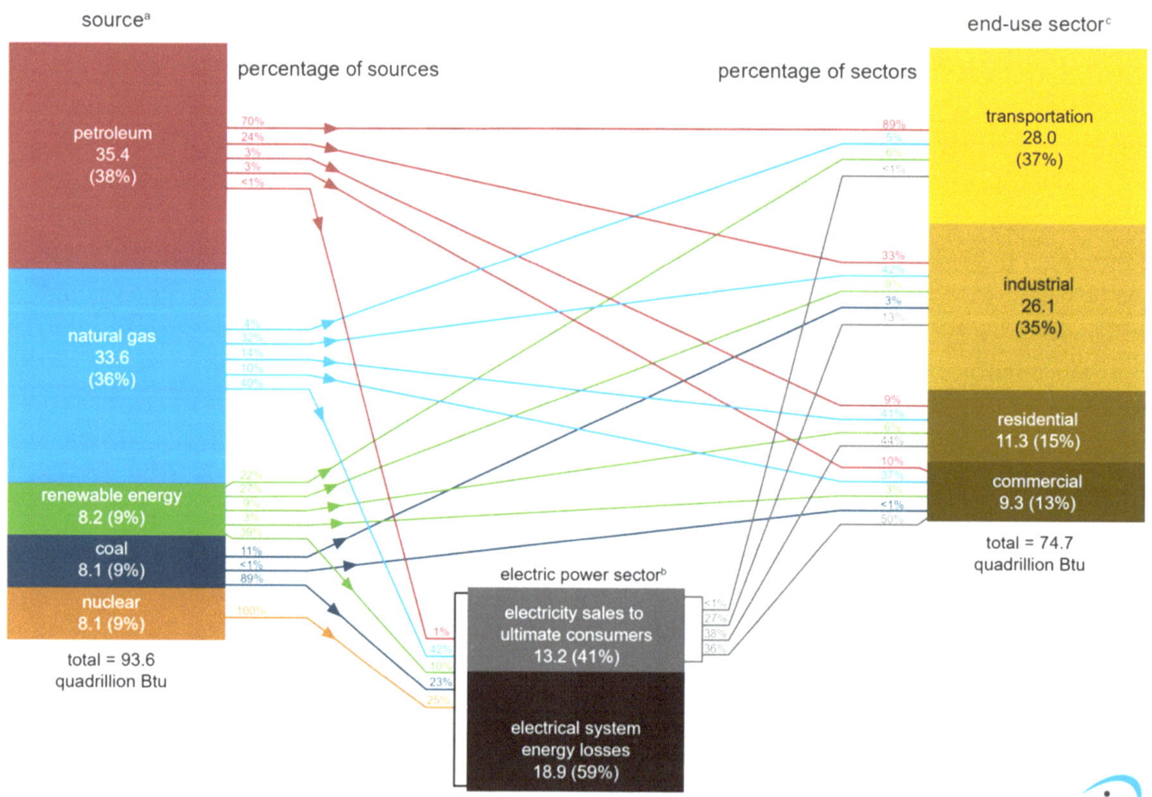

Reprinted from Ref. [4].

Transportation Sector

Petroleum, primarily used for transportation, holds the largest proportion at 37% [1.5, 1.6]. Energy consumption within the transportation sector includes the energy used by automobiles; trucks; buses; motorcycles; trains, subways, and other rail vehicles; aircraft; and ships, barges, and other waterborne vehicles whose primary purpose is to transport people and/or goods from one physical location to another [1.7]. Vehicles whose primary purpose is not transportation (e.g., construction cranes and bulldozers, farming vehicles, and warehouse forklifts) are classified in the sector of their primary use. The energy sources for this entire proportion are fossil-derived liquid fuels. In order to meet this need, it is necessary to keep the reserves under control and ensure resource security in the regions with oil across the world.

Distillation of crude oil and making its by-products suitable for use by industries make up another large proportion of energy use. Petroleum products are obtained from the processing of crude oil, natural gas, and other hydrocarbon compounds [1.8]. Such products include unfinished oils, hydrocarbon gas liquids, aviation gasoline, motor gasoline, jet fuel, kerosene, distillate fuel oil, residual fuel oil, petrochemical feedstocks, special naphthas, lubricants, waxes, petroleum coke, asphalt, road oil, still gas, and miscellaneous products [1.9]. Energy is also consumed in the production, transfer, and storage of these products. The fact that oil reserves are located in a small number of countries and the available reserves are limited makes it difficult to supply raw materials for petroleum-consuming sectors [1.10]. The countries with the highest production of petroleum products in 2023 are shown in **Figure 1.3** [1.11]. In addition to their petroleum production capacity, the energy consumption of these countries is also quite high. The production of fuel products that are especially needed in the transportation sector is of great social and economic importance. Since the transportation sector ranks first in this consumption, petroleum products must be produced, inspected, and shipped in accordance with all vehicles in the sector [1.12].

The gradual increase in the number of vehicles in the transportation sector has further increased the dependence on petroleum-derived liquid fuels. The widespread use of petroleum-derived fuels has led to the rapid depletion of limited petroleum resources, higher global greenhouse gas (GHG) emissions, and the release of pollutants such as carbon monoxide (CO), particulate matter (PM), and hydrocarbon (HC) emissions [1.13]. Transportation is also one of the most significant contributors to global climate change. In fact, total GHG emissions originating from the transportation sector, which is dependent on oil, hover around 14% [1.14]. The sources and distribution of GHG emissions originating from the transportation sector indicate that a large proportion of emissions comes from both passenger and freight transportation on the road [1.15].

Figure 1.3 Countries with the highest production of petroleum products in 2023 [1.11].

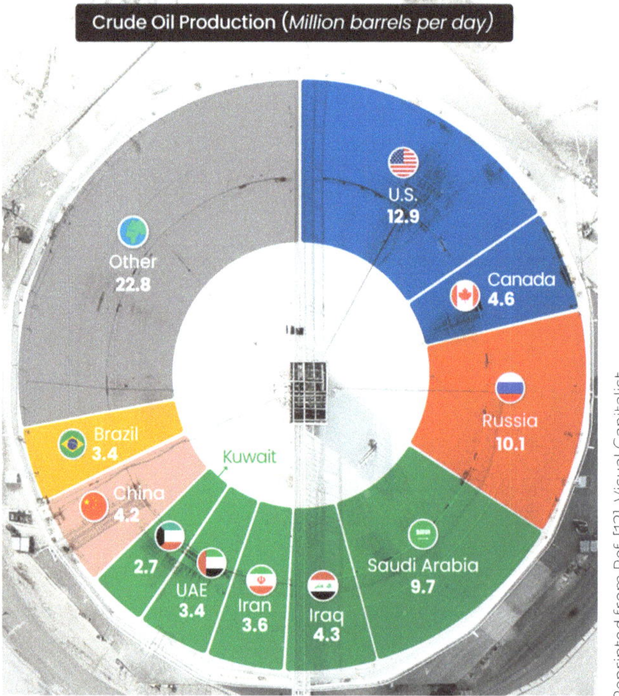

Reprinted from Ref. [12]. Visual Capitalist.

Piston-type engines working with liquid fuels are used in ground transportation vehicles. Except for the defense sector, internal combustion engines (ICEs) are designed to achieve smaller volumes and higher performance. These play into the indispensable elements of the transportation sector of saving both time and fuel and gaining commercial benefits [1.16]. However, the emergence of GHGs in both passenger and freight transportation in the transportation sector has necessitated serious measures be taken for the utilization of such vehicles in the sector [1.17]. Goals of reaching zero emissions, in an effort to reduce global climate change, are driving the world to cut all sources of GHGs, requiring efforts for emission reductions across multiple sectors.

Due to the intense use of passenger cars in social life, a series of goals have already been taken by policymakers in terms of both community and environmental health. Primary efforts include the goal of decarbonizing ground transportation to a large extent by 2050, in line with the GHG emissions from various passenger car powertrains and fuels, and the Paris Agreement targets [1.18]. Accomplishing such goals will require the implementation of a variety of measures, such as the use of electric vehicles (an example projection is shown in **Figure 1.4**).

Figure 1.4 Lifecycle GHG emissions of passenger cars, 2030 projection [1.19].

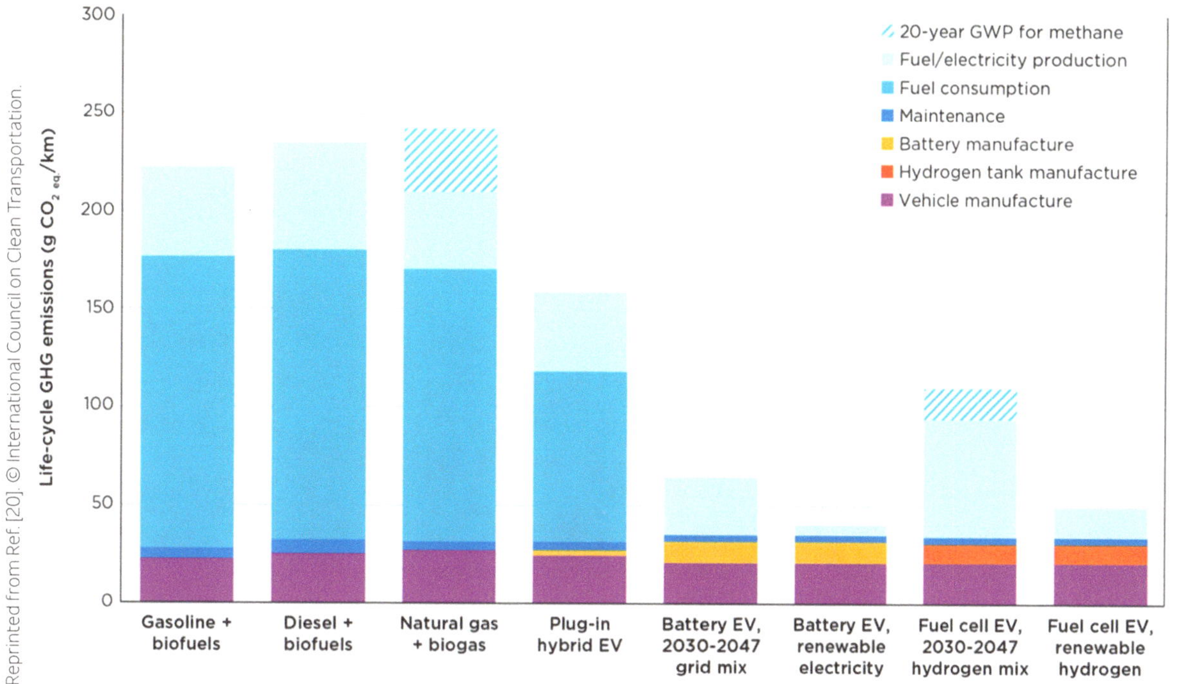

In order to reduce GHG emissions, biofuel-based vehicles are important in the short term, while vehicles powered by battery electric (BE), fuel cells (FCs), and electrosynthetic fuels (e-fuel) play a bigger role in the long term. Historically, the first way to achieve low emissions was the use of biofuels, and at the end of long-term research, fuels produced from biomass were used in both diesel and gasoline engines [1.19]. However, due to the increasing number of cars and limited oil reserves, the concept of zero carbon has begun to dominate. In this context, the US Environmental Protection Agency (EPA) proposes new, more stringent emissions standards for medium- and heavy-duty trucks by 2032. The EPA "projects the 2027-2032 model year rules would cut more than 9 billion tons of CO_2 emissions through 2055 [1.20]." In recent years, direct electrification has also become a popular solution for the decarbonization of transportation. Automotive manufacturers are expected to adopt electric power instead of petroleum-derived fuel power as the driving force in the future [1.18, 1.19, 1.20].

Air Transportation

Airplanes, another means of public and freight transportation, have shown significant growth over time. **Figure 1.5** tracks the number of passengers who used air transportation from 1945 to 2022 worldwide, and it is seen that there is only an increasing demand, excluding the COVID-19 pandemic period [1.21].

Figure 1.5 Worldwide air passenger traffic between 1945 and 2022 [1.21].

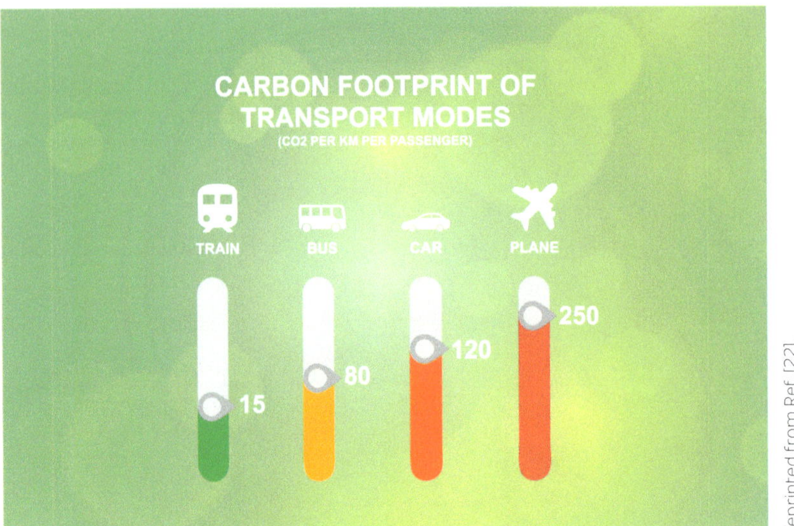

Reprinted from Ref. [22].

Global passenger traffic recovered modestly in 2021, based on the latest International Civil Aviation Organization (ICAO) economic impact analysis of COVID-19 on civil aviation, revealing that the number of passengers worldwide was 2.3 billion, or 49% below pre-pandemic (2019) levels, up from the 60% drop observed in 2020 [1.21, 1.22]. Considering the time and speed factors together, air transport will need continuous development both technically and economically to meet future demands.

Air transportation, as well as ground transportation, contributes a large proportion of GHG emissions. Aviation-based emissions increased in 2021, rebounding to a level between their pre-pandemic peak in 2019 and the level in 2020 [1.22]. After increasing at an average of 2.3% per year from 1990 to 2019, the COVID-19 pandemic led to the plummeting of direct CO_2 emissions from aviation from more than 1000 Mt in 2019 to 600 Mt in 2020. Emissions in 2021 totaled around 720 Mt CO_2, regaining nearly one-third of the fall seen in the previous year. They are expected to grow rapidly, surpassing their 2019 level in the next few years [1.23]. Passenger transport efficiency (passenger-kilometers) by mode of transport shows an increasing trend in aviation, as shown in **Figure 1.6** [1.15].

Figure 1.6 Passenger transport activity (passenger-kilometers) by mode [1.15].

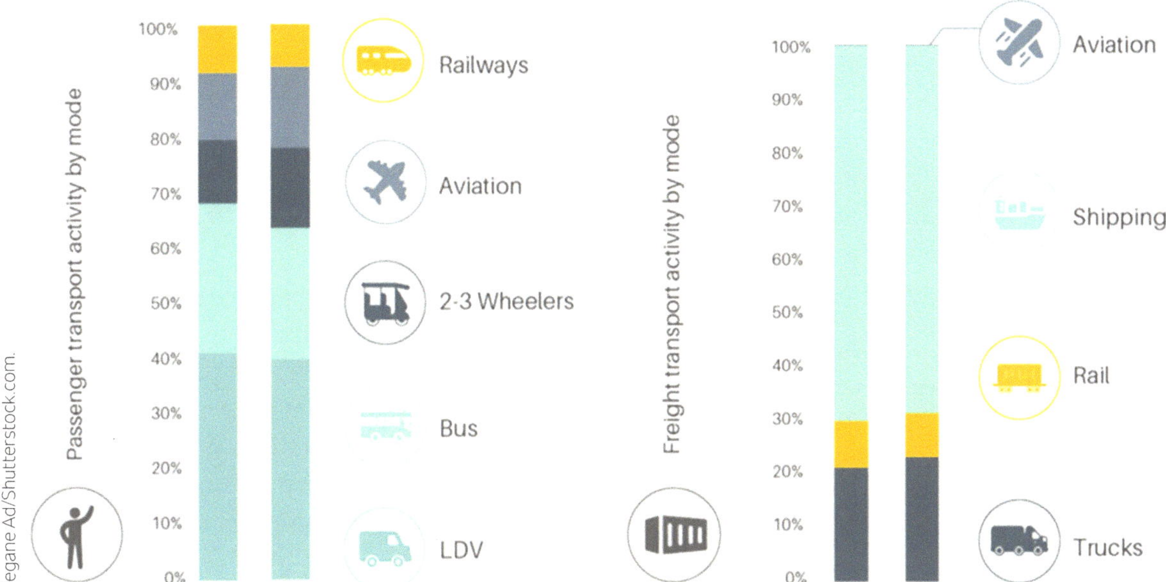

Passenger demand gradually recovered in 2021, with domestic traffic at 68% of 2019 levels and international traffic at just 28%. This represented an overall increase of 28% compared to 2020. The 2024 global total passenger count is close to 9.5 billion, a 9% increase compared to 2023 and a 3.8% increase compared to pre-pandemic levels (2019). The ten busiest airports, representing 9% of global commercial air traffic (855 million passengers), saw an 8.8% increase compared to 2023 and an 8.4% increase compared to 2019 (789 million passengers in 2019). Air cargo, however, showed stronger growth in 2021, increasing by nearly 7% above the pre-pandemic peak [1.24, 1.25]. Air cargo volumes at the top ten airports increased by 9.3% annually in 2024 and by 9.5% compared to 2019 (29.5 million tons). When the daily number of flights of airline companies and the duration of their stay in the air are examined, the necessity of controlling the pollutants released into the atmosphere becomes ever starker.

Power Generation, Liquid Fuel, and Pollution of Aviation

The air transportation industry will continue to be heavily dependent on ICEs, such as turboprop, turbofan, and turbojet, for power generation because of their limited requirements for energy density, operational safety, and systematic reliability [1.26]. As shown in **Figure 1.7**, piston-type engines and gas turbines are widely used in the aviation industry.

Due to the thermodynamic working principle, a liquid fuel should be used in these engines [1.27]. The liquid fuel requirement is obtained by the distillation of petroleum fuels. The use of these engines in the aviation industry significantly contributes to global climate change [1.28]. In addition to CO_2 emissions, pollutants such as HC, CO, and nitrogen oxide (NO_x) increase according to the type of engine and fuel used [1.29]. It has become a necessity to keep the levels of all these pollutants under control in terms of both human and environmental health and to reach zero emissions in the long term. At this point, the effect of aviation on climate change should be handled carefully, and every effort should be made to remove the influence of the consequences of its use. While fuel efficiency, gas turbine technological innovation, and improved flight logistics will improve matters, they will fall significantly short of the emission reductions needed to truly combat climate change [1.29]. Multiple additional measures will be needed to promote innovative technologies, operations, and market conditions necessary to achieve low emissions [1.25, 1.29].

When it comes to taking precautions on airplanes, companies take a step back in terms of cost, especially flight safety. In parallel with developments in material technology, lighter aircraft have been built, which have had a positive impact on flight performance. New aircraft are up to 20% more efficient than the models they replace, but this has been insufficient to keep up with the growing activity [1.30]. Between 2000 and 2010, fuel efficiency improved by 2.4% per annum, and between 2010 and 2019, it improved by 1.9%, demonstrating that additional incremental improvements are becoming more difficult [1.31, 1.32]. Fuel is one of the most important factors determining the transportation cost in the aviation sector, where the majority of cargo transfers are made, especially passenger transportation by air.

Figure 1.7 Aircraft engine types.

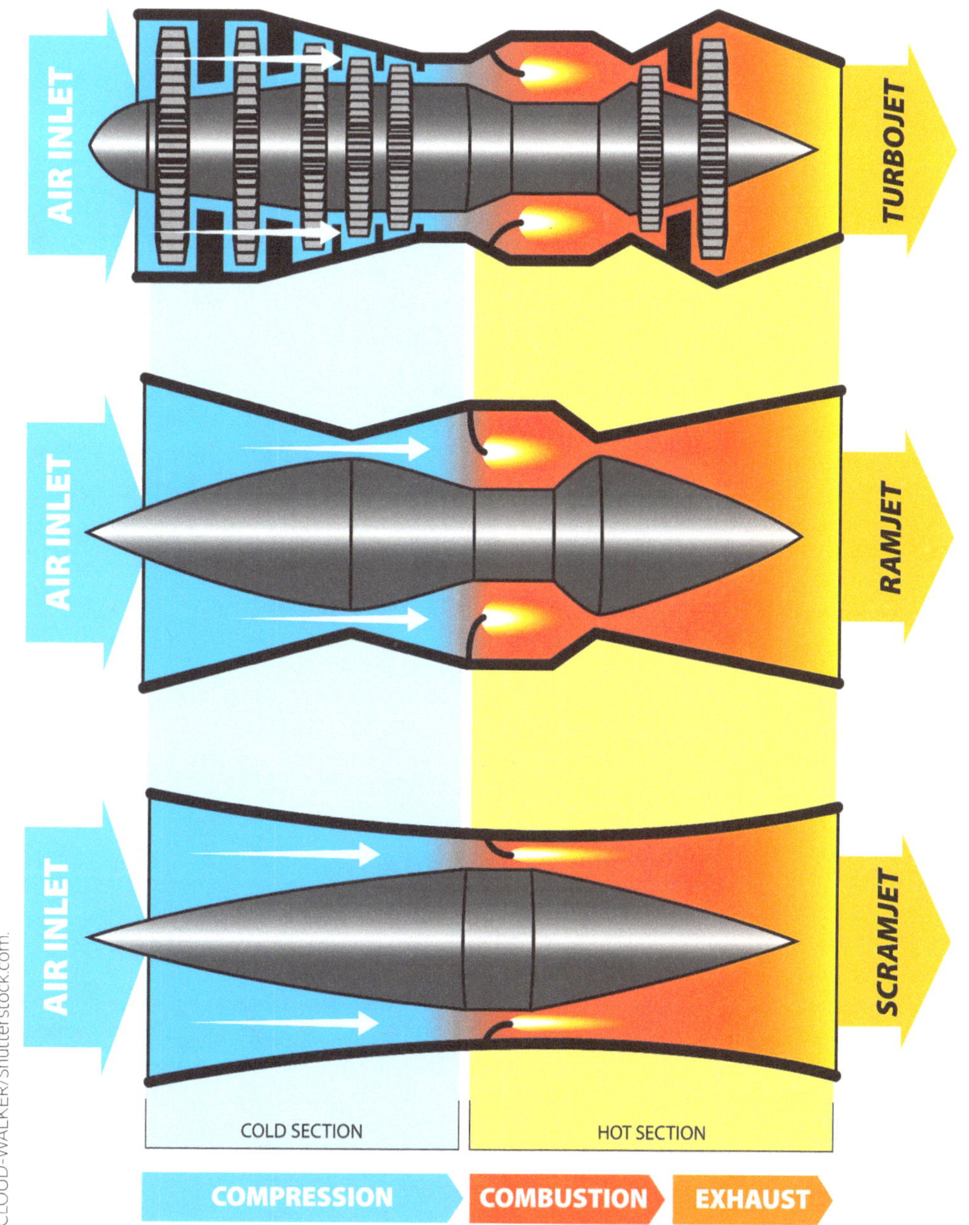

Aviation Fuel Market

Conventional aviation fuels are petroleum-based fuels or petroleum and synthetic fuel blends used to power aircraft. Aviation fuels have more stringent requirements than those used in land vehicles and contain additives to enhance or maintain properties important to fuel performance or handling. While kerosene-based jet propellant (JP-B and Jet A-1) is used in gas turbine-powered aircraft, piston-based aircraft use leaded gasoline or jet fuel (kerosene) [1.32]. The economic mobility of aviation fuels throughout the world is increasing every year. The increasing need for fuels used in commercial aircraft (narrow-body aircraft and wide-body aircraft), regional jets, business jets, helicopters, and military aircraft in the sector affects this economic evaluation. The market size of aviation fuels was valued at around $238.2 billion in 2024 and will exhibit a growth rate of more than 8% compound annual growth rate (CAGR) from 2025 to 2034, driven by the increasing technological advancements and investments for optimizing fuel formulation techniques and minimizing environmental impacts [1.33]. The sector size is expected to hit 3.17 billion barrels in volume with an anticipated CAGR of 8.2% through the forecasted period. Thus, the aviation fuel market, from its current market value of over $300 billion, is predicted to increase to more than $474.9 billion by 2034, as shown in **Figure 1.8** [1.33]. The significantly higher growth rate is attributed to the continued development and investment in aviation fuels to power future aircraft.

Figure 1.8 Market value of aviation fuels worldwide [1.33].

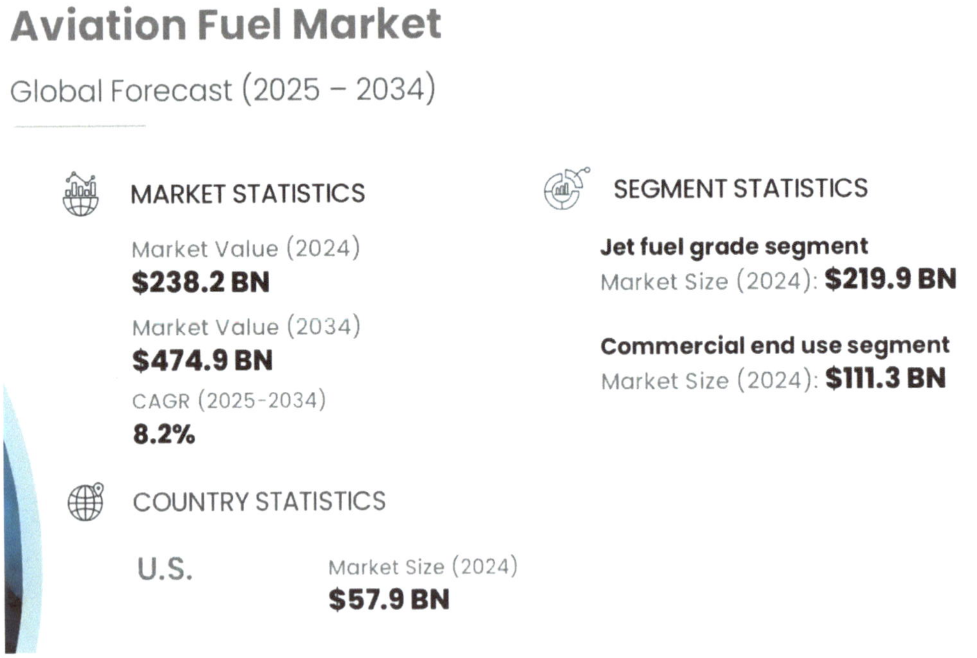

The Importance of SAFs

Environmental factors in air transportation indicate development and associated goals. However, aviation engines often use heavy fossil fuels, which are characterized by high carbon emissions and unsustainable development due to their chemical structure [1.34]. To achieve a global carbon neutralization target by 2050, it is necessary to find new alternative aviation fuels for airplanes to mitigate the zero-carbon problem in both the general and civil aviation sectors [1.35]. Aircraft using petroleum-derived fuels contribute to increased atmospheric pollutants. Studies have been conducted for many years to eliminate these issues in road vehicles by using alternative fuels, and countries encourage using biofuels to protect the environment and reduce oil dependence [1.36]. Therefore, there is an urgent need for environmentally friendly and sustainable energy sources. Sustainable energy sources will continue to grow as long as economically feasible, especially if government subsidies are available to support their production.

Promoting efficiency through financial and regulatory measures is among the most pressing near- and medium-term priorities. It encourages investment in sustainable fuels and developing alternatives to jet kerosene, such as BE, e-fuel, and hydrogen-powered airplanes [1.37]. Although electric vehicles have entered the future projection of ground transportation after many years, the use of battery technology and electric motors in aircraft used in other transportation systems does not seem likely in the short term. The energy density of today's Li-ion batteries is about 200 Wh/kg at the cell level, but for short-haul flights over 1000 km, a battery pack energy density of at least 800 Wh/kg would be required. Furthermore, the lower energy storage capability of batteries compared to HC fuels has so far prevented electric aircraft from using battery technologies [1.38]. For these reasons, the current battery energy density and weight severely limit the range of BE-powered flights and the size of the aircraft.

The most promising approach, then, for the aviation sector is to use alternative fuels with chemical structures that can exhibit properties close to those of petroleum-derived fuels. This will require first the creation of sustainable-biofuel-powered aircraft and then the widespread adoption of advanced technologies for SAFs [1.39]. To make the aviation industry cleaner and greener, SAF is no longer just an idea but it is a necessity for low-carbon air travel. Sustainable biofuels produced from biomass (residue raw materials) and renewable waste resources are used all over the world as an alternative to fossil fuels, which can reduce pollutant emissions and increase energy security for the energy sector, and it is time to bring such alternative fuels to the aviation industry. The current SAF cycle is presented in **Figure 1.9** [1.40].

Figure 1.9 Fueling cycle for SAFs [1.40].

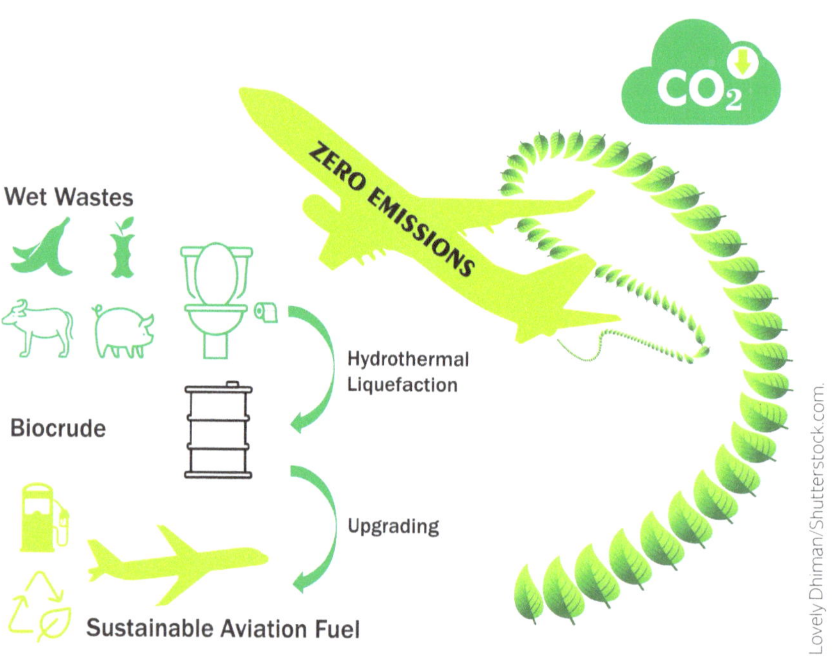

Depending on the feedstocks and technologies used to produce the fuel, SAFs can reduce lifecycle GHG emissions dramatically compared to conventional jet fuel. Over the lifecycle, SAF reduces GHG emissions by up to 80% compared to fossil-based aviation fuels [1.40].

The production process and pathways of aviation biofuels are easily applicable using existing technologies [1.41]. As shown in **Figure 1.10**, SAFs can be produced from a wide range of feedstocks using five key processing pathways, including oil to jet, alcohol to jet (AtJ), gas to jet, and sugar to jet, among others [1.42]. Fuels such as SAFs have the advantage that a few or no modifications are necessary on the aircraft itself, provided that fuel characteristics meet the specifications for lubricity and density as well as adequately swelling elastomer seals in current aircraft fuel systems. Thus, SAFs can be used as a direct replacement (drop-in) for fossil jet fuel as they are chemically similar. They are fully compatible with existing jet engines and fueling infrastructure, requiring zero additional investment. The SAF market is segmented by technology: Fischer–Tropsch (FT) fuels, hydro-processed esters and fatty acids (HEFAs), catalytic hydrothermolysis jet (CHJ), hydropro-cessed fermented sugars to synthesized isoparaffins (HFS-SIPs), and AtJ application (commercial and defense) [1.43].

Figure 1.10 SAF production pathways [1.42].

SAFs can be produced from a wide range of feedstocks using 5 key processing pathways

Mapping of feedstocks to pathways

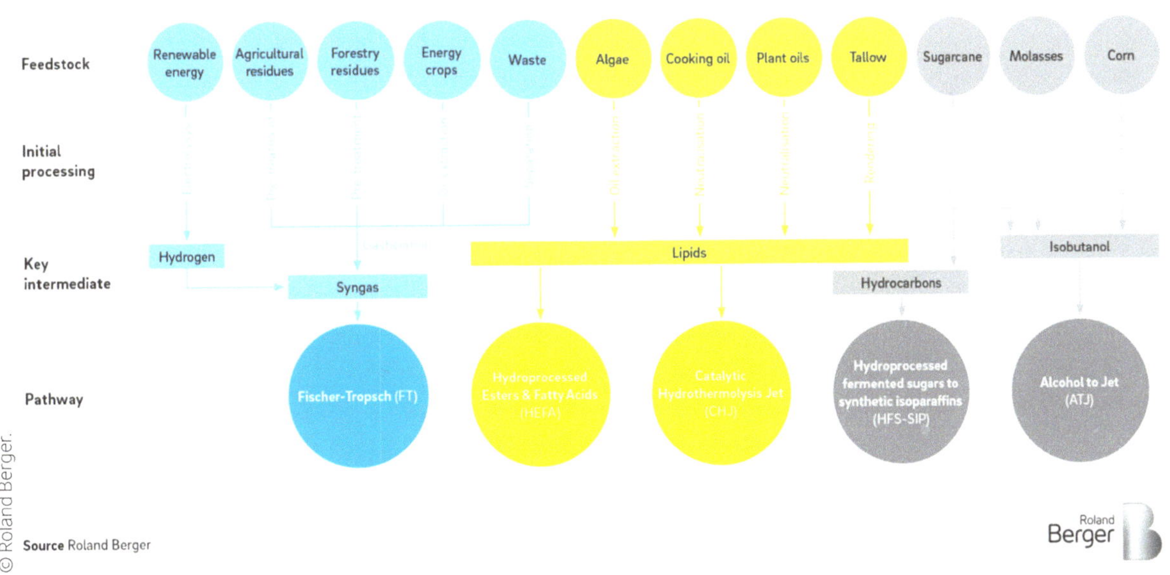

Moreover, the initial phase of these efforts involves the widespread adoption of biofuels in the transition to electric vehicles, followed by mandating the use of biofuels for other vehicles, such as aircraft and ships that rely on liquid fuels. The increasing demand for sustainable and clean energy sources provides incentives for the development of alternative fuels [1.44]. Simultaneously, the development of gas turbine technologies with a flexible fuel supply system enables the use of alternative non-fossil fuels that can play a key role in contributing to global efforts to meet emissions targets. SAF is a cleaner, direct replacement for fossil jet fuel [1.45].

More than 450,000 commercial flights have used SAFs since 2016; more than 50 airlines and 13 major airports currently use and supply SAFs; and their numbers are growing rapidly. SAFs are projected to supply approximately 0.1 to 0.15% of total aviation fuel in 2022 [1.46]. SAFs can be mixed up to 50% with conventional jet fuel, and all quality tests meet conventional jet fuel standards. Thus, any aircraft certified for using the current specification of jet fuel can use SAFs. The demand for SAFs is also increasing, and a greater proportion of renewable energy has the potential to solve the problems of energy scarcity and environmental pollution. The global market for SAFs was valued at 186.6 million dollars in

2021. It is estimated to advance at a CAGR of 26.2% from 2022 to 2050 and reach 402 billion dollars by the end of 2050. By geography, the market is segmented into North America, Europe, Asia Pacific (APAC), South America, and Middle East and Africa (MENA) [1.47].

Current Policies

In addition to social and economic development, it is crucial to implement sustainable policies in the aviation sector to protect the environment. Achieving a sustainability revolution in aviation requires decisive action from organizations such as the ICAO and the International Air Transport Association (IATA), where both developed and developing countries collaborate. At this point, constructive policies are needed to support SAF consumption, which is necessary to achieve economic targets and accelerate demand growth [1.48]. The use of SAFs is increasing globally, particularly in Europe, the United Kingdom (UK), and the US, as governments and industry work together to find ways to steadily decarbonize the aviation sector through government fuel subsidies. Including Brazil's RENOVABIO (certificate trading), California's Low Carbon Fuel Standard (LCFS, based on carbon density), ReFuelEU Aviation (a blending instruction) proposed by the EU, and the Sustainable Aviation Fuel Grand Challenge proposed by the US, a number of regulatory frameworks have been or are being implemented [1.48, 1.49, 1.50]. Many airlines are partnering with suppliers to achieve their SAF goals. However, to transition more effectively to sustainable fuels in the aviation sector, the current use of SAFs should be better supported and promoted. Additionally, to ensure the successful introduction of new SAF production routes with high emission reduction potential, it is necessary to adopt an optimum SAF mixing ratio to create a long-term consistent support structure. Key policy developments that can assist with this goal include the US Sustainable Aviation Fuel Grand Challenge, which includes funding to develop fuel and aircraft technologies, and the ReFuelEU Aviation regulation proposed to determine blending. ReFuelEU includes an obligation to aviation fuel suppliers to blend a minimum proportion of SAFs with fossil jet kerosene, with the aim of achieving 63% by 2050 from 2% in 2025 [1.49, 1.50]. The commercial sale of SAFs is subject to blending limits, but with 100% SAF flights completed and further trials ongoing, the sale of 100% SAFs is expected by 2050 [1.48, 1.49]. Within the framework of the policies adopted and accepted to reduce carbon emissions worldwide by 2050, countries have actions and future policy target times as of 2024. **Figure 1.11** illustrates the targeted projections for reducing carbon emissions by country. Accordingly, the aviation sector will also receive its proportion of these targets, and the steps to be taken for the use of SAF are included in these projections.

IATA has already taken steps in the right direction by committing to achieving carbon-neutral growth from 2020 onward and net-zero carbon emissions from the global air transport industry operations by 2050 [1.52, 1.53]. Unlike passenger cars, the jet fleet currently cannot be fully electrified with battery technologies. Accordingly, the development and full implementation of SAFs will continue to be important into the 21st century.

Figure 1.11 Targeted projections for reducing carbon emissions by country [1.51].

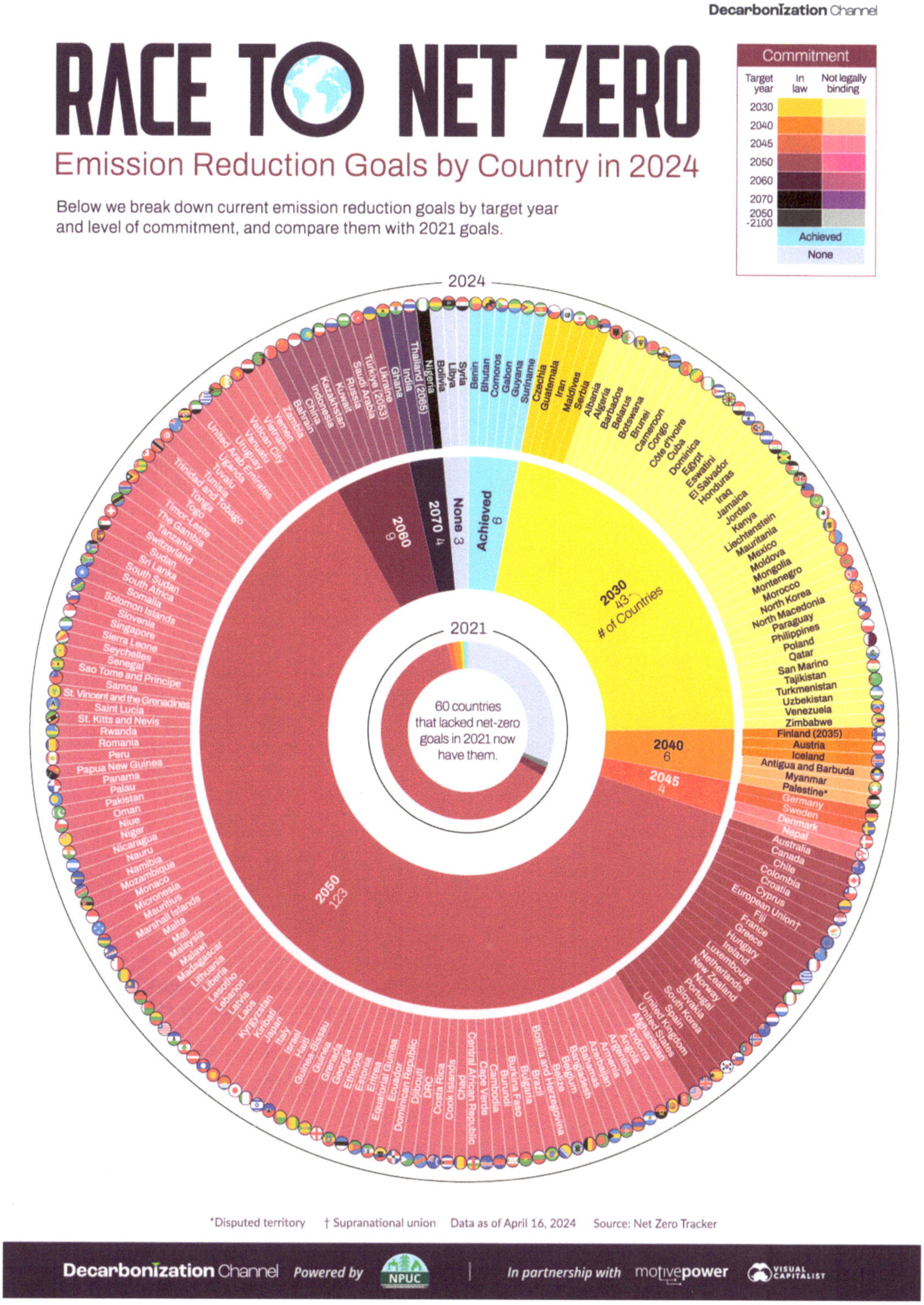

Decarbonization Channel
RACE TO NET ZERO
Emission Reduction Goals by Country in 2024
Below we break down current emission reduction goals by target year and level of commitment, and compare them with 2021 goals.
Commitment
Target year
In law
Not legally binding
2030
2040
2045
2050
2060
2070
2050 -2100
Achieved
None
2024
2021
60 countries that lacked net-zero goals in 2021 now have them.
2030
43
# of Countries
2040
6
2045
4
2050
123
2060
9
2070
4
None 3
Achieved
6
Benin
Bhutan
Comoros
Gabon
Guyana
Suriname
Czechia
Guatemala
Iran
Maldives
Serbia
Albania
Algeria
Barbados
Belarus
Botswana
Brunei
Cameroon
Congo
Côte d'Ivoire
Cuba
Dominica
Egypt
Eswatini
El Salvador
Honduras
Iraq
Jamaica
Jordan
Kenya
Liechtenstein
Mauritania
Mexico
Moldova
Mongolia
Montenegro
Morocco
North Korea
North Macedonia
Paraguay
Philippines
Poland
Qatar
San Marino
Tajikistan
Turkmenistan
Uzbekistan
Venezuela
Zimbabwe
Finland (2035)
Austria
Iceland
Antigua and Barbuda
Myanmar
Palestine*
Germany
Sweden
Denmark
Nepal
Australia
Canada
Chile
Colombia
Croatia
Cyprus
European Union†
Fiji
France
Greece
Hungary
Ireland
Luxembourg
Netherlands
New Zealand
Norway
Portugal
Slovakia
South Korea
Spain
United Kingdom
United States
Afghanistan
Andorra
Angola
Argentina
Armenia
Azerbaijan
Bahamas
Bangladesh
Belgium
Belize
Bosnia and Herzegovina
Brazil
Bulgaria
Burkina Faso
Burundi
Cambodia
Cape Verde
Central African Republic
Chad
Cook Islands
Costa Rica
DRC
Djibouti
Dominican Republic
Ecuador
Equatorial Guinea
Eritrea
Estonia
Ethiopia
Georgia
Grenada
Guinea
Guinea-Bissau
Haiti
Israel
Italy
Japan
Kiribati
Kyrgyzstan
Laos
Latvia
Lebanon
Lesotho
Liberia
Lithuania
Madagascar
Malawi
Malaysia
Mali
Malta
Marshall Islands
Mauritius
Micronesia
Monaco
Mozambique
Namibia
Nauru
Nicaragua
Niger
Niue
Oman
Pakistan
Palau
Panama
Papua New Guinea
Peru
Romania
Rwanda
St. Kitts and Nevis
Saint Lucia
St. Vincent and the Grenadines
Samoa
Sao Tome and Principe
Senegal
Seychelles
Sierra Leone
Singapore
Slovenia
Solomon Islands
Somalia
South Africa
South Sudan
Sri Lanka
Sudan
Switzerland
Tanzania
The Gambia
Timor-Leste
Togo
Tonga
Trinidad and Tobago
Tunisia
Tuvalu
Uganda
United Arab Emirates
Uruguay
Vanuatu
Vatican City
Vietnam
Yemen
Zambia
Bahrain
China
Indonesia
Kazakhstan
Kuwait
Russia
Saudi Arabia
Türkiye (2053)
Ukraine
Ghana
India
Thailand (2065)
Nigeria
Bolivia
Libya
Syria
*Disputed territory
† Supranational union
Data as of April 16, 2024
Source: Net Zero Tracker
Decarbonization Channel Powered by NPUC
In partnership with motivepower
VISUAL CAPITALIST

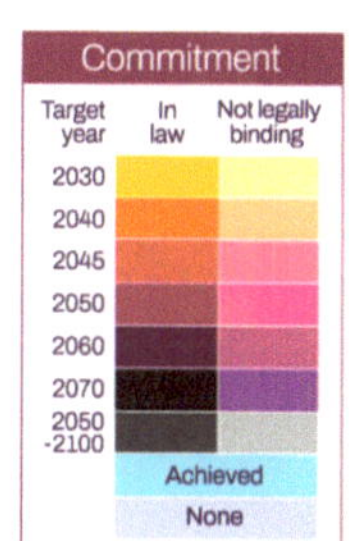

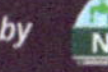

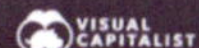

Reprinted from Ref. [52]. Visual Capitalist.

References

1.1. Solangi, Y.A. and Magazzino, C., "Evaluating Financial Implications of Renewable Energy for Climate Action and Sustainable Development Goals," *Renewable and Sustainable Energy Reviews* 212 (2025): 115390, doi:https://doi.org/10.1016/j.rser.2025.115390.

1.2. Ali, K., Jianguo, D., and Kirikkaleli, D., "How Do Energy Resources and Financial Development Cause Environmental Sustainability?" *Energy Rep* 9 (2023): 4036-4048, doi:https://doi.org/10.1016/j.egyr.2023.03.040.

1.3. U.S. Energy Information Administration (EIA), "U.S. Energy Facts Explained," accessed January 20, 2025, https://www.eia.gov/energyexplained/us-energy-facts/.

1.4. U.S. Energy Information Administration (EIA), "Monthly Energy Review," accessed February 5, 2025, https://www.eia.gov/totalenergy/data/monthly/.

1.5. BP, "Energy Outlook 2024 Edition," accessed February 8, 2025, https://www.bp.com/content/dam/bp/business-sites/en/global/corporate/pdfs/energy-economics/energy-outlook/bp-energy-outlook-2024.pdf.

1.6. BP, "Sustainability Report 2024," accessed February 8, 2025, https://www.bp.com/content/dam/bp/business-sites/en/global/corporate/pdfs/sustainability/group-reports/bp-sustainability-report-2024.pdf.

1.7. Jiang, F., Ma, L., Broyd, T., Li, J. et al., "Systematic Framework for Sustainable Urban Road Alignment Planning," *Transport Res D: Transport Environ* 120 (2023): 103796, doi:https://doi.org/10.1016/j.trd.2023.103796.

1.8. Blokhina, P.M., Molchanov, E.E., Znoyko, S.A., Usoltsev, S.D. et al., "Development of a New Environmentally Friendly Method for Determining Aromatic Compounds in Light Petroleum Products," *Optical Materials* 160 (2025): 116746, doi:https://doi.org/10.1016/j.optmat.2025.116746.

1.9. Jia, C., Pang, X., and Song, Y., "Basic Principles of the Whole Petroleum System," *Petroleum Exploration and Development* 51, no. 4 (2024): 780-794, doi:https://doi.org/10.1016/S1876-3804(24)60506-9.

1.10. Zhen, X., Vinnem, J.E., Han, Y., Peng, C. et al., "Development and Prospects of Major Accident Indicators in the Offshore Petroleum Sector," *Process Safety and Environmental Protection* 160 (2022): 551-562, doi:https://doi.org/10.1016/j.psep.2022.02.050.

1.11. Venditti, B., "The World's Biggest Oil Producers in 2023," accessed January 5, 2025, https://www.visualcapitalist.com/the-worlds-biggest-oil-producers-in-2023/.

1.12. Geissler, C.H., Ryu, J., and Maravelias, C.T., "The Future of Biofuels in the United States Transportation Sector," *Renewable and Sustainable Energy Reviews* 192 (2024): 114276, doi:https://doi.org/10.1016/j.rser.2023.114276.

1.13. Yilmaz, N. and Davis, S.M., "Diesel Blends with High Concentrations of Biodiesel and n-Butanol: Effects on Regulated Pollutants and Polycyclic Aromatic Hydrocarbons," *Process Safety and Environmental Protection* 166 (2022): 430-439, doi:https://doi.org/10.1016/j.psep.2022.08.041.

1.14. Olivari, E., Caballini, C., and Lluch, X., "How to Calculate GHG Emissions in Freight Transport? A Review of the Main Existing Online Tools," *Case Studies on Transport Policy* 19 (2025): 101343, doi:https://doi.org/10.1016/j.cstp.2024.101343.

1.15. SLOCAT Partnership on Sustainable, Low Carbon Transport, "Global Transport and Climate Change," accessed October 23, 2024, https://tcc-gsr.com/global-overview/global-transport-and-climate-change/.

1.16. Hasan, M.A., Chapman, R., and Frame, D.J., "Acceptability of Transport Emissions Reduction Policies: A Multi-Criteria Analysis," *Renew. Sustain. Energy Rev.* 133 (2020): 110298, doi:https://doi.org/10.1016/j.rser.2020.110298.

1.17. Dehdari, P., Wlcek, H., and Furmans, K., "An Updated Literature Review of CO_2e Calculation in Road Freight Transportation," *Multimodal Transport* 2 (2023): 100068, doi:https://doi.org/10.1016/j.multra.2022.100068.

1.18. Linton, C., Grant-Muller, S., and Gale, W.F., "Approaches and Techniques for Modelling CO_2 Emissions from Road Transport," *Transp. Rev.* 35 (2015): 533-553, doi:https://doi.org/10.1080/01441647.2015.1030004.

1.19. Searle, S., Bieker, G., and Baldino, C., "Decarbonizing Road Transport by 2050: Zero-Emission Pathways for Passenger Vehicles," accessed November 21, 2024, https://theicct.org/publication/decarbonizing-road-transport-by-2050-zero-emission-pathways-for-passenger-vehicles/.

1.20. U.S. Environmental Protection Agency (EPA), "Proposed Rule: Multi-Pollutant Emissions Standards for Model Years 2027 and Later Light-Duty and Medium-Duty Vehicles," accessed August 12, 2024, https://www.epa.gov/regulations-emissions-vehicles-and-engines/proposed-rule-multi-pollutant-emissions-standards-model.

1.21. The International Civil Aviation Organization (ICAO) Newsroom, "2021 Global Air Passenger Totals Show Improvement from 2020, But Still Only Half Pre-Pandemic Levels," December 7, 2024, https://www.icao.int/Newsroom/Pages/2021-global-air-passenger-totals-show-improvement.aspx.

1.22. Javanmard, M.E., Tang, Y., and Martínez-Hernández, J.A., "Forecasting Air Transportation Demand and Its Impacts on Energy Consumption and Emission," *Applied Energy* 364 (2024): 123031, doi: https://doi.org/10.1016/j.apenergy.2024.123031.

1.23. Cavallaro, F. and Nocera, S., "COVID-19 Effects on Transport-Related Air Pollutants: Insights, Evaluations, and Policy Perspectives," *Transport Reviews* 44, no. 2 (2024): 484-517, doi:https://doi.org/10.1080/01441647.2023.2225211.

1.24. Jing, H., Chen, Y., Ma, M., Feng, W. et al., "Global Carbon Transition in the Passenger Transportation Sector over 2000–2021," *Sustainable Production and Consumption* 51 (2024): 556-571, doi:https://doi.org/10.1016/j.spc.2024.10.006.

1.25. Airports Council International (ACI), "The Busiest Airports in the World Defy Global Uncertainty and Hold Top Rankings," April 16, 2025, https://aci.aero/2025/04/14/the-busiest-airports-in-the-world-defy-global-uncertainty-and-hold-top-rankings/.

1.26. The International Air Transport Association (IATA), "Air Cargo Market Analysis," accessed December 12, 2024, https://www.iata.org/en/iata-repository/publications/economic-reports/air-cargo-market-analysis-december-2024/.

1.27. Chen, R., Jiang, C., Wang, K., and Yang, H., "Engine Option in Aircraft Purchase: Company Strategies and Policy Implications," *Transport Policy* 159 (2024): 266-283, doi:https://doi.org/10.1016/j.tranpol.2024.10.011.

1.28. U.S. National Aeronautics and Space Administration (NASA), "Types of Gas Turbines," accessed August 12, 2024, https://www.grc.nasa.gov/www/k-12/airplane/trbtyp.html.

1.29. Climate Change, "Calculating the True Climate Impact of Aviation Emissions," accessed September 12, 2024, https://www.climatechange.ie/guest-post-calculating-the-true-climate-impact-of-aviation-emissions/.

1.30. The International Civil Aviation Organization (ICAO) Environmental Protection, "Aircraft Engine Emissions," accessed August 21, 2024, https://www.icao.int/environmental-protection/Pages/default.aspx.

1.31. Ovaere, M. and Proost, S., "Strategic Climate Policy in Global Aviation: Aviation Fuel Taxes and Efficiency Standards with Duopolistic Aircraft Producers," *Economics of Transportation* 41 (2025): 100397, doi:https://doi.org/10.1016/j.ecotra.2025.100397.

1.32. Ryerson, M.S. and Kim, H., "The Impact of Airline Mergers and Hub Reorganization on Aviation Fuel Consumption," *Journal of Cleaner Production* 85 (2014): 395-440, doi:https://doi.org/10.1016/j.jclepro.2013.12.032.

1.33. Inan, I., Orhan, I., and Ekici, S., "Fuel Savings Strategies for Sustainable Aviation in Accordance with United Nations Sustainable Development Goals (UN SDGs)," *Energy* 320 (2025): 135159, doi:https://doi.org/10.1016/j.energy.2025.135159.

1.34. Global Market Insight, "Aviation Fuel Market Size – By Grade, End Use, Application Analysis, Share, Growth Forecast, 2025–2034," accessed March 3, 2025, https://www.gminsights.com/industry-analysis/aviation-fuel-market.

1.35. Shao, L., Zhou, Y., Geng, T., Zhao, S. et al., "Advanced Combustion in Heavy Fuel Aircraft Piston Engines: A Comprehensive Review and Future Directions," *Fuel* 370 (2024): 131771, doi:https://doi.org/10.1016/j.fuel.2024.131771.

1.36. Jensen, L.L., Bonnefoy, P.A., Hileman, J.I., and Fitzgerald, J.T., "The Carbon Dioxide Challenge Facing U.S. Aviation and Paths to Achieve Net Zero Emissions by 2050," *Progress in Aerospace Sciences* 141 (2023): 100921, doi:https://doi.org/10.1016/j.paerosci.2023.100921.

1.37. Vardon, D.R., Sherbacow, B.J., Guan, K., Heyne, J.S. et al., "Realizing 'Net-Zero-Carbon' Sustainable Aviation Fuel," *Joule* 6, no. 1 (2022): 16-21, doi:https://doi.org/10.1016/j.joule.2021.12.013.

1.38. Bardon, P., Massol, O., and Thomas, A., "Greening Aviation with Sustainable Aviation Fuels: Insights from Decarbonization Scenarios," *Journal of Environmental Management* 374 (2025): 123943, doi:https://doi.org/10.1016/j.jenvman.2024.123943.

1.39. Bai, M., Yang, W., Yan, J., Zhang, R. et al., "Cryogenic Turbo-Electric Hybrid Propulsion System with Liquid Hydrogen Cooling for a Regional Aircraft," *International Journal of Hydrogen Energy* 71 (2024): 541-561, doi:https://doi.org/10.1016/j.ijhydene.2024.05.161.

1.40. Waddington, E.G. and Ansell, P.J., "Reprint of: A Definition, Conceptual Framework, and Pathway towards Sustainable Aviation," *Progress in Aerospace Sciences* 153 (2025): 101083, doi:https://doi.org/10.1016/j.paerosci.2025.101083.

1.41. BP, "What Is Sustainable Aviation Fuel (SAF)?," accessed November 12, 2024, https://www.bp.com/en/global/air-bp/news-and-views/views/what-is-sustainable-aviation-fuel-saf-and-why-is-it-important.html.

1.42. Arias, A., Nika, C.-E., Vasilaki, V., Feijoo, G. et al., "Assessing the Future Prospects of Emerging Technologies for Shipping and Aviation Biofuels: A Critical Review," *Renewable and Sustainable Energy Reviews* 197 (2024): 114427, doi:https://doi.org/10.1016/j.rser.2024.114427.

1.43. Roland Berger, "Sustainable Aviation Fuels Key for the Future of Air Travel," accessed December 2, 2024, https://www.rolandberger.com/en/Insights/Publications/Sustainable-aviation-fuels-key-for-the-future-of-air-travel.html.

1.44. Su-ungkavatin, P., Tiruta-Barna, L., and Hamelin, L., "Biofuels, Electrofuels, Electric or Hydrogen?: A Review of Current and Emerging Sustainable Aviation Systems," *Progress in Energy and Combustion Science* 96 (2023): 101073, doi:https://doi.org/10.1016/j.pecs.2023.101073.

1.45. Karimi, M., Simsek, H., and Kheiralipour, K., "Advanced Biofuel Production: A Comprehensive Techno-Economic Review of Pathways and Costs," *Energy Conversion and Management: X* 25 (2025): 100863, doi:https://doi.org/10.1016/j.ecmx.2024.100863.

1.46. BP, "Sustainable Aviation Fuel–Key Facts," accessed December 2, 2024, https://www.bp.com/en/global/air-bp/aviation-fuel/sustainable-aviation-fuel.html.

1.47. I6 Group, "Which Airlines Are Embracing SAF?," accessed September 3, 2024, https://www.i6.io/blog/which-airlines-are-embracing-saf.

1.48. Transparency Market Research, "Sustainable Aviation Fuel Market," accessed November 12, 2024, https://www.transparencymarketresearch.com/sustainable-aviation-fuel-market.html.

1.49. The International Civil Aviation Organization (ICAO) Environment, "Sustainable Aviation Fuels (SAF)," accessed August 5, 2024, https://www.icao.int/environmental-protection/pages/SAF.aspx.

1.50. European Commission, "ReFuelEU Aviation," accessed September 15, 2024, https://transport.ec.europa.eu/transport-modes/air/environment/refueleu-aviation_en.

1.51. U.S. Department of Energy, "Synthetic Aviation Fuel Grand Challenge," accessed September 15, 2024, https://www.energy.gov/eere/bioenergy/synthetic-aviation-fuel-grand-challenge.

1.52. Visual Capitalist, "Visualized: Emission Reduction Targets by Country in 2024," accessed September 16, 2024, https://www.visualcapitalist.com/sp/visualized-emission-reduction-targets-by-country-in-2024/.

1.53. The International Air Transport Association (IATA), "Net Zero Roadmaps," accessed September 20, 2024, https://www.iata.org/en/programs/sustainability/roadmaps/.

Chapter 02

Production Pathways to SAFs

With the global energy demand expected to increase by 48% over the next 20 years due to rapid population growth, currently 80% of energy demand is fulfilled by fossil fuels. However, the rapid depletion of fossil fuel reserves and the negative environmental impacts resulting from the combustion of fossil fuels have increased the demand for sustainable biofuels. This has become a major factor in the transition to a carbon-neutral bioeconomy in developed countries [2.1]. Aviation is an important part of the domestic and international transportation and cargo transportation network. As the need for aviation fuels increases, the demand for cleaner fuel supply systems and renewable resources has also increased. It is of great importance in terms of aviation safety that the global availability and quality of the fuel are at uniform standards worldwide [2.2]. Fluctuation in oil prices is the main driver of pricing in the aviation industry. Aviation is one of the sectors that contributes the highest to total GHG emissions. In this context, the number of flights and their duration have led to a sustainable fuel policy. The primary method to reduce pollutants from ICEs and dependence on fossil-derived fuels is the use of sustainable biofuels. Energy crops that can be used as production sources must be inedible in order not to affect food production and the ecosystem, including the environment, so that it does not result in deforestation [2.3]. The use of alternative aviation fuels produced from various sources has become a promising approach to reducing harmful emissions from aviation and ensuring energy sustainability [2.4]. Thus, sustainable biofuels should offer low carbon emissions throughout their lifecycle. Moreover, SAFs should be capable of being used without any changes to the infrastructure of today's aircraft engines and fuel systems. Developing sustainable biofuels and diversifying resources will support the production of high-quality fuels for the aviation industry. Worldwide, biomass production is approximately 100 trillion kilograms annually [2.5]. It is clear that biomass is a potential resource that can replace fossil fuels when economical and efficient production methods are developed.

Using alternative jet fuels produced from non-conventional sources has become an important strategy in achieving sustainable aviation. Technologies such as thermochemical and biochemical methods allow us to produce aviation biofuels from biomass resources. In this context, SAF feedstocks, production methods, and technologies will be discussed in this section.

Feedstocks

Biofuels and biofuel blends, classified under SAFs, will be crucial in achieving net-zero future. The aviation industry has set a target of net-zero carbon emissions by 2050 for IATA member airlines and has implemented regulations in every field to achieve this [2.6]. New technologies seek to change the future of aviation by providing feasible ways to reduce GHG emissions in the aviation sector, which is one of the most difficult sectors to decarbonize. For biofuels to be viable alternatives to fossil-derived aviation fuels, the collection of feedstocks, production pathways, properties, and characterization of the produced fuels should be considered comprehensively. In general, "biofuels" refers to fuels produced using biological resources (plant or animal material). Overall, there is a growing demand for biofuels worldwide. The US, Europe, Brazil, and Indonesia are responsible for the majority of the progress in biodiesel, renewable diesel, and SAF consumption. As shown in **Figure 2.1**, the demand for these fuels is estimated to increase by 44% or 21 billion liters over the period of 2022–2027 [2.7].

Figure 2.1 Total biofuel production growth by region, 2021–2027 [2.7].

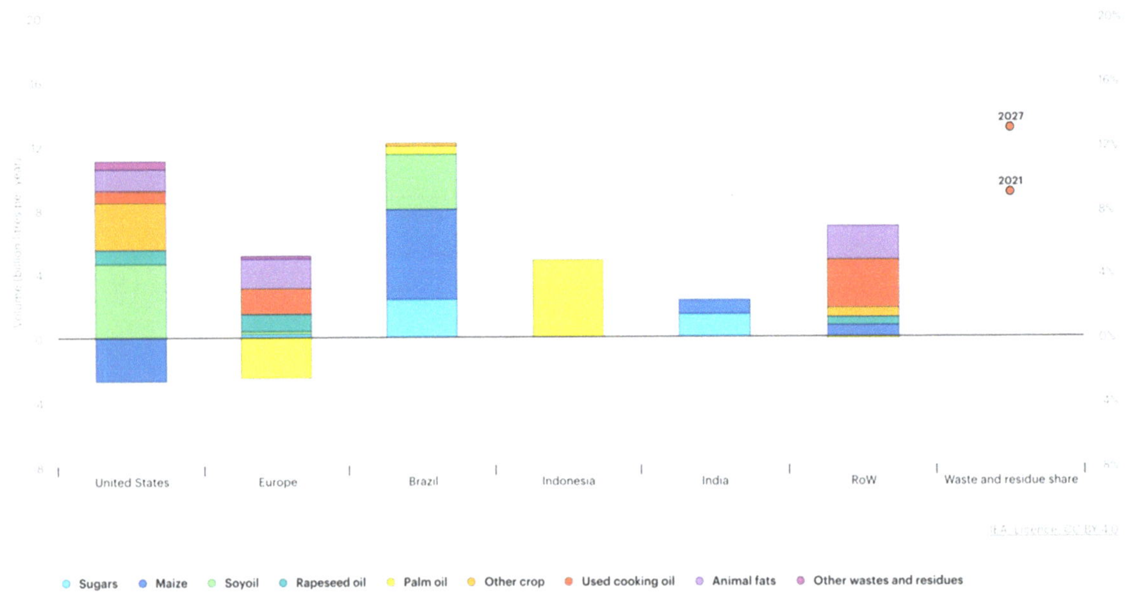

SAFs constitute a significant proportion of these fuels, which have also benefited from the rate of production growth. SAFs can be produced from both fossil and bioresources. However, aviation fuels that can be produced from sustainable bioresources have become even more important. SAFs can be produced from a variety of feedstocks such as oil seed plants and energy grasses; agricultural and forestry residue; organic municipal solid waste (MSW); fats, oils, and greases (FOG) from cooking waste and meat production; algae; and industrial CO waste gas (as shown in **Figure 2.2**) [2.7, 2.8].

Figure 2.2 Feedstocks suitable for SAF production [2.8].

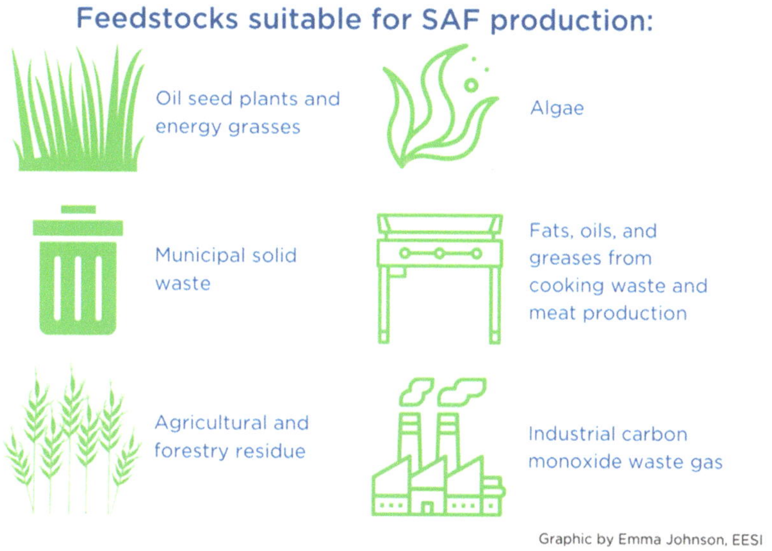

The feedstocks to be used in renewable fuel production should meet basic criteria such as being easily obtainable, sustainable, and inedible (non-food energy crops) [2.9]. The use of energy plants (such as algae, *Camelina*, and *Jatropha*) in the supply of raw materials and waste that can be used in fuel production varies depending on the region. Among them, inedible vegetable and animal oils and all kinds of oil-containing waste have a great potential as sources suitable for being converted to SAFs [2.10]. In this case, SAFs will replace traditional jet fuel, allowing the use of multiple products from various raw materials and production technologies. Although SAFs are an excellent alternative to conventional aviation fuels, the collection of raw materials, the production route, and the characteristics and characterization of the fuel produced are issues that need to be addressed. The properties and content of the feedstock directly affect the method by which the fuel is produced. In addition to the various production processes, the raw materials used also play an important role. EU Directive 2009/28/EC (Renewable Energy Directive) defines and regulates the use of raw materials for biofuels in Europe, including aviation [2.11]. In the current literature and industry, feedstocks used in biofuel production are classified into first generation to fourth generation, depending on the raw material and/or biosynthetic platform (i.e., genetic engineering). Feedstocks that can be used in biofuel production have been classified according to their use and production processes over the years, as shown in **Figure 2.3** [2.12].

Figure 2.3 Feedstock generation for SAF production [2.12].

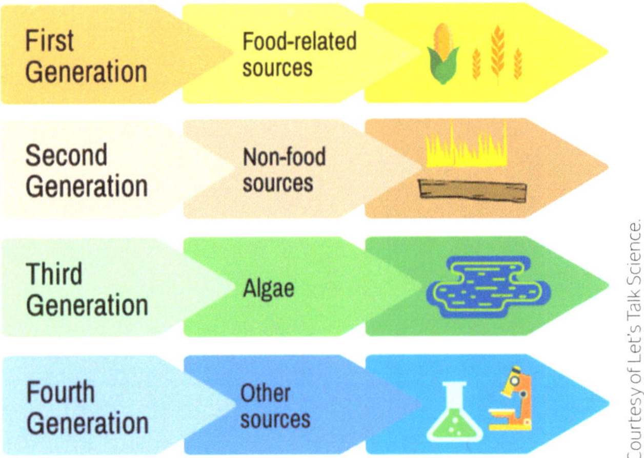

Courtesy of Let's Talk Science.

Although SAFs can be produced from various feedstocks, their availability in terms of volume and location is limited. Additionally, there are concerns about competition with food sources, high carbon footprints, and land waste [2.13]. The main factors in choosing these feedstocks include sustainability, biomass development, biomass breakdown, and conversion process. When evaluated in terms of sustainability, biomass is easily available and its production should not have a negative impact on the environment. In addition, applying growth techniques that will increase the oil content of biomass and using production techniques to obtain maximum oil yield are highly important factors in biofuel production [2.14]. The generation of biofuels is defined according to the generation of the feedstocks used. **Figure 2.4** shows the classification of biofuels by raw material type [2.15].

Figure 2.4 Biofuel generations according to feedstocks [2.15].

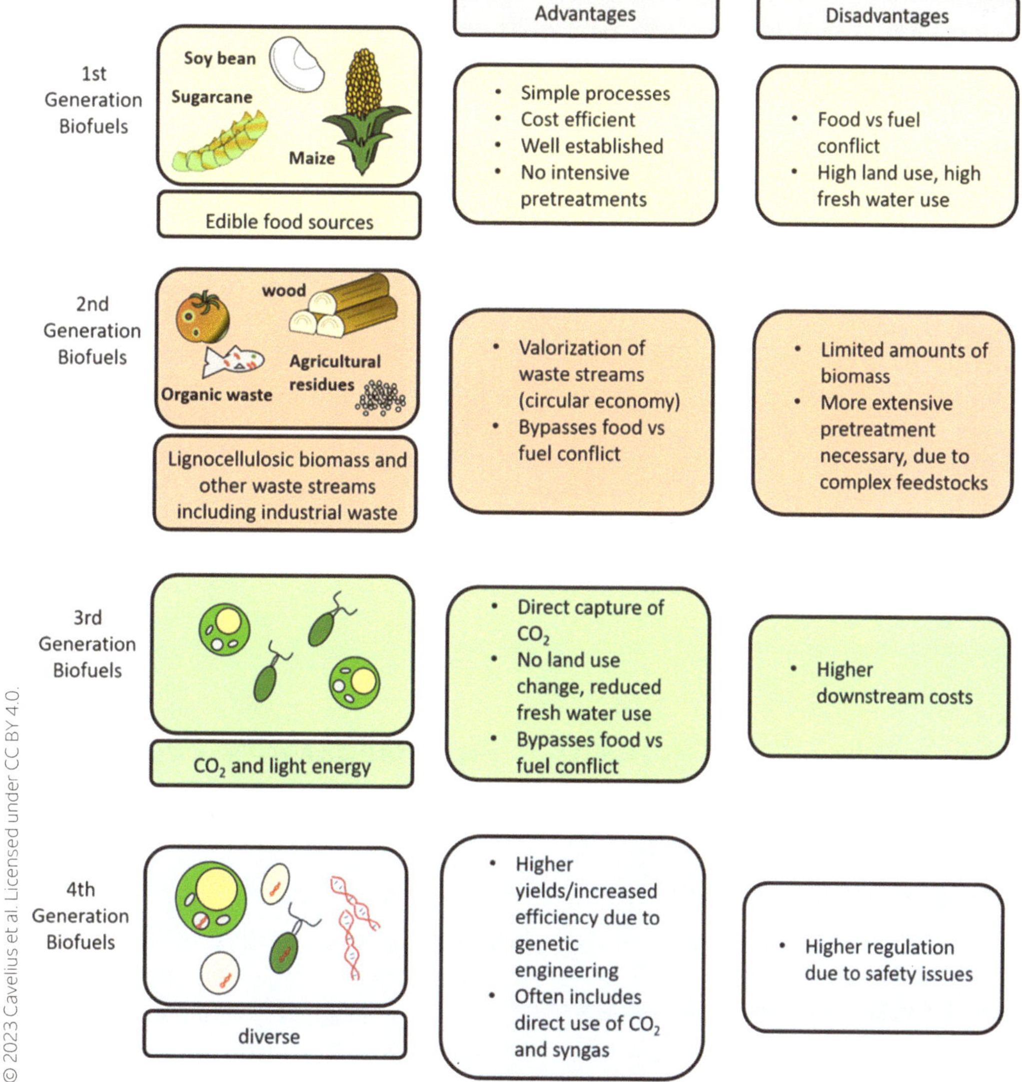

As shown in **Figure 2.4**, some advanced and waste feedstocks have higher sustainability potential than others. While the demand for SAFs is driven primarily by North America and Europe, most of the feedstocks available today are grown in Asia. Therefore, manufacturers and end users must consider investing in the availability of feedstocks [2.16]. Feedstocks that can be used in SAF production around the world have been categorized by ICAO Carbon Offsetting and Reduction Scheme for International Aviation (CORSIA) to ensure the safety and standardization of SAFs in the short term. Additionally, within the framework of CORSIA, feedstocks used to produce SAFs are generally divided into five categories as shown in **Table 2.1** [2.17].

Table 2.1 SAF feedstock categories under the CORSIA framework [2.17].

Feedstock	Classification	Feedstock	Classification
Palm fatty acid distillate	By-product	Processing residues: crude glycerine	Residue
Technical corn oil		Processing residues: crude tall oil	
Molasses	Co-product	Processing residues: empty palm fruit bunches	
Brassica carinata oil	Main product	Processing residues: forestry processing residues	
Camelina oil		Processing residues: palm oil mill effluent	
Corn grain		Processing residues: sewage sludge	
Jatropha oil		Processing residues: tall oil pitch	
Miscanthus (herbaceous energy crops)		Agricultural residues: bagasse cobs, husks, manure, nut, shells, stalks stover, and straw	
Palm oil		Forestry residues: bark branches, cutter shavings, leaves, needles, precommercial thinning, slash, and tree tops	
Poplar (short-rotation woody crops)		Municipal solid waste (MSW)	Waste
Rapeseed oil		Tallow	
Soybean oil		Used cooking oil	
Sugar beet		Waste gases	
Sugarcane			
Switchgrass (herbaceous energy crops)			

Primary and by-products are the main products of a production process. These products have significant economic value and flexible supply. By-products are secondary products whose supply is inelastic with economic value. Waste is materials whose supply is inelastic with no economic value. This definition does not include raw materials or substances that have been deliberately altered or contaminated to meet this criterion. Residues are secondary materials whose supply is inelastic and of little economic value. Under the CORSIA framework, raw materials in the categories of by-products, wastes, and residues are eligible for an indirect land-use change (ILUC) value of zero in calculating the lifecycle emission value of the SAF [2.18].

First-Generation Feedstocks

First-generation feedstocks consist of high-oil-content plants and seeds that can also be used for food or animal feed. Maize, wheat, grain, sugarcane, sugar beet, rapeseed, palm, soybeans, waste oil, and animal fat are some examples of feedstocks that are widely used in many parts of the world [2.19]. Low-carbon alcohols can be produced using microorganisms and enzymes through the fermentation of sugars, starches, or cellulose. Among these alcohols, bioethanol has the highest production potential. Bioethanol and biodiesel can be produced in the easiest way from these feedstocks. Bioethanol production can be easily achieved via fermentation using biochemical methods [2.20]. Similarly, biodiesel can be produced using thermochemical methods. Bioethanol is one of the prominent SAFs. The demand for SAFs continues to grow, but the aviation industry is challenged by limited supplies of traditional SAF raw materials such as vegetable oils, animal fats, and waste oils [2.21]. Additionally, bioethanol is used as an additive in combination with petroleum. Ethanol has replaced lead because it acts as an octane improver. The combination of petroleum (90%) and bioethanol (10%) is the most common mixture in the fossil fuel industry [2.22]. However, since this generation of raw materials is edible, there is a risk of depletion of resources used in the food industry if they are used in fuel production. In addition to food security, the fact that agricultural lands have become extremely inefficient due to first-generation feedstocks grown for fuel production can lead to environmental degradation. In addition, most of the biofuels produced from first-generation feedstocks are dependent on subsidies and are not comparable with respect to cost with existing fossil fuels such as oil, and some biofuels provide only limited GHG emission savings. Moreover, the production of first-generation feedstocks negatively affects GHG emissions, biodiversity, land use, water use, and water pollution due to increased fertilizer use, resulting in increasing amounts of nitrogen and phosphorus entering soil and surface waters. When emissions from production and transportation are considered, the lifecycle assessment (LCA) of first-generation biofuels often exceeds that of conventional fossil fuels [2.23].

For these reasons, although these raw materials can be produced very easily, especially in Europe and the US, the need to diversify fuel sources in a way that does not affect food safety has emerged [2.24]. Feedstocks must also meet sustainability criteria, such as lifecycle carbon emissions, for the term "sustainable" to be validly used. For this reason, the production of first-generation feedstocks using freshwater resources causes problems such as the inability to compete with the necessary food production and deforestation. Therefore, sustainable raw materials required for SAFs must be met by waste feedstocks.

Second-Generation Feedstocks

Feedstock compositions and production processes directly influence biofuel properties. Second-generation feedstocks have been developed to overcome the limitations of primitively produced biofuels. They consist of parts of a biological resource or plants and wastes with oil content that are not directly edible [2.25]. The negative effects of the first-generation feedstocks on food security have been eliminated in the second generation, but it is a raw material generation whose potential in terms of agricultural land use needs to be determined. Second-generation feedstocks consist of perennial energy plants (willow, poplar), short-rotation forest crops, agricultural residues, forestry residues, and inedible crops of oil plants [2.26]. The International Energy Agency (IEA) has identified non-food biomass sources such as "lingo-cellulosic materials, including cereal straw, bagasse, forest residues and lingo-cellulosic materials, including purpose-grown energy crops such as grasses and short-rotation forests" as potential feedstocks [2.27].

Great progress has been made in the development of second-generation biofuels, particularly using lignocellulosic waste biomass obtained from agricultural and forestry waste and other wastes (e.g., animal fats or those from the food industry such as waste cooking and frying oil). Using these resources in biofuel production increases both environmental and agricultural efficiency. There are advantages in using inevitable by-products of the agricultural industry in biofuel production. The most important advantage is that no additional fertilizer, water, or land is required to grow the feedstock. Some of the by-products, especially inedible by-products, are used to produce animal feed, but a significant amount can also be used in biofuel production. The large-scale utilization and application of lignocellulose feedstocks remains a crucial initiative to provide sufficient alternative fuels in the global biofuels market [2.28]. The process of producing biofuels from second-generation feedstocks does not require high capital cost to maintain and operate sophisticated equipment. The feedstock required for their production can be grown on low-quality marginal land with low GHG emissions and low conversion efficiency. In this respect, it has also created an economic potential for energy production as shown in **Figure 2.5** [2.29]. In the near future, ethanol, which can be produced from second-generation feedstocks using advanced technologies, will play a major role in achieving the 2050 target. Ethanol remains an important alternative fuel for ICEs in biofuel production. The fact that it can be produced from second-generation feedstocks ensures its continued availability and sustainability.

Figure 2.5 Ethanol production from second-generation feedstocks [2.29].

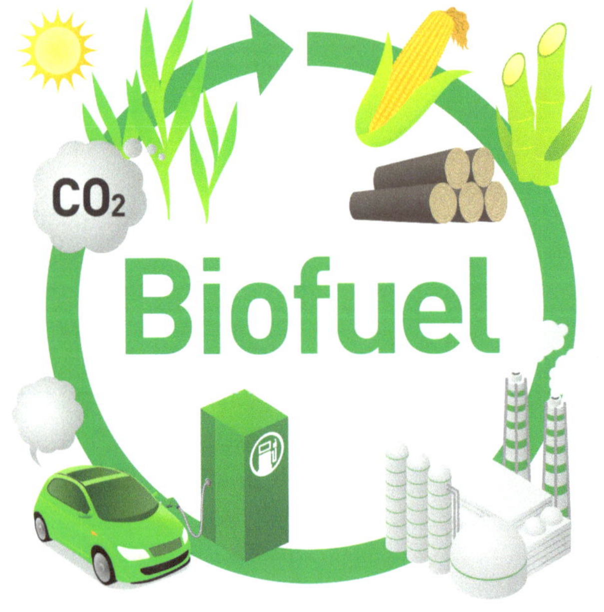

It is anticipated that there will be a new generation of SAF conversion technologies that can utilize existing bioproducts such as ethanol, which has the potential for rapid production, or convert biomass derived from sources such as forest waste, fuel crops, MSW, and advanced alcohols (e.g., cellulosic ethanol) into fuels [2.30, 2.31]. These second-generation feedstocks could address many of the concerns facing first-generation biofuels and offer the potential for further cost reductions in the long term [2.29, 2.30].

Third-Generation Feedstocks

Biofuels produced from second-generation feedstocks overcome the main challenges associated with first-generation biofuels, but there are still some challenges. Second-generation biofuels require large amounts of land for sustainable production, which has economic implications in terms of high agricultural costs [2.32]. Furthermore, the availability of cellulosic material from second-generation feedstocks can be regionally insufficient. Therefore, in terms of production potential, third-generation feedstocks in the form of algal-based biofuels have the potential to address many of the shortcomings of first- and second-generation biofuels [2.33]. Third-generation biofuels from algae are now scientifically recognized as excellent biofuel alternatives. Microalgae, important feedstocks for third-generation biofuels, can grow in seawater, wastewater, food waste, and even saline alkaline soil, and have the potential to produce biomass at a higher rate than terrestrial plants using limited nutrients [2.33, 2.34]. The development of microalgae as an alternative fuel feedstock is of great importance in advancing energy conversion, alleviating food pressure, environmental protection, and economic value [2.35]. Biofuel production pathways from algae-based feedstocks continue to be developed to be more environmentally friendly and economical over large production areas. Global commercial production and consumption of liquid, solid, and gaseous biofuels that can be produced from third-generation feedstocks continue to grow, despite their short-term economic inefficiencies. Third-generation biofuels have high potential for growth worldwide as the biofuel market continues to expand every day.

The US Department of Energy estimates that mass-producing algae-based biofuels could meet all fuel needs of the US using only 0.42% of the country's land area. It also predicts that the production cost of algae-based biofuel will be much higher than that of fuel from other sources due to the additional needs of algae production. More research and development is needed to mass-produce algae without major environmental consequences [2.36, 2.37]. Today, the development of advanced processes for large-scale production is in the early stage, and large quantities are not yet common. However, there are promising advantages in terms of higher non-edible oil content, improved efficiency of land used in production, and higher availability of the resulting oil. SAF production pathways from algae-based feedstocks will continue to emerge to scale cleaner energy technologies for the wider economy [2.35, 2.36, 2.37]. In order to produce SAFs from algae and compete with conventional jet fuel, it must first be produced commercially on a large scale. While test flight studies are available, as shown in **Figure 2.6**, scaling up depends on several factors that are not currently available, including further development of third-generation SAF production infrastructure with state-of-the-art production processes and American Society for Testing and Materials (ASTM) approval for increasing SAF blending limits [2.38].

Figure 2.6 Using algae-based SAFs [2.38].

However, there are still outstanding issues affecting the overall production and market opportunities worldwide, thus deterring potential investors for algae production. Overall, third-generation biofuels are still in the research and development phase compared to first- and second-generation biofuels.

Fourth-Generation Feedstocks

Fourth-generation feedstocks are the newest type of feedstock. These feedstocks include genetically modified algae and cyanobacteria that produce high levels of oil. They do not compete with edible foods, no arable or fertile land is required to grow them, and their growth rates are much faster [2.39]. The basic principle of fourth-generation biofuels is the conversion of available solar energy into useful inedible oil using feedstocks. Fourth-generation biofuels aim to use genetically optimized feedstocks designed to increase the capture of carbon dioxide (CO_2) [2.40]. While biofuels from third-generation feedstocks are still being discussed, a fourth generation is emerging in the bioenergy community. Biofuels from fourth-generation feedstocks will include genetically modified biomass crops that will produce high yields and require little horticulture. As shown in **Figure 2.7**, they are often referred to as an extension of third-generation feedstocks [2.41]. Microalgae that have been genetically modified and grown in a laboratory environment can have high oil content, using which larger quantities of raw materials can be produced faster. However, in today's technology, it is important to reduce the costs of genetic modification. Fuels produced from these feedstocks are still under research at the laboratory level. Synthetic biology and biomass processing methods can help develop fourth-generation feedstocks by increasing photon-fuel efficiency. The initial investment required to produce biofuels from this feedstock is quite high [2.42].

Figure 2.7 Third- and fourth-generation feedstocks [2.41].

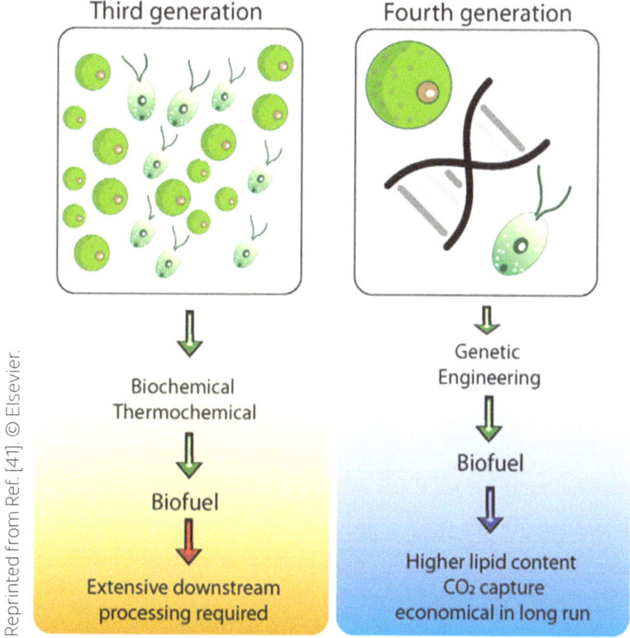

In recent years, research and development on fourth-generation feedstocks has been advancing due to their easy availability and high energy content. However, there has been no mass production of SAFs from this generation of feedstocks in terms of commercialization potential [2.43]. Current feedstocks used in the production of advanced biofuels require more pretreatment procedures and involve higher costs. At this point, the feasible sequestration technology required to ensure the sustainability of the feedstock has limited the commercialization of this process.

Feedstock Availability and Sustainability

In selecting the suitable type of feedstock for SAF production, extensive research has been conducted on different biofuel feedstock generations and potential alternatives [2.44]. However, no alternative exists without prohibitive consequences. Hence, a combination of biofuel generations could be a viable option for commercialization. Continuous research and development is needed to find an alternative feedstock that is renewable, is sustainable, and has no major economic or environmental disadvantages in fuel production [2.45]. Additionally, ICAO CORSIA's studies on the sustainability of feedstocks are still ongoing. Current requirements focus solely on GHG emissions and carbon stock. While this is a positive first step, focus is also needed on feedstock supply to enable fuels to meet higher sustainability targets [2.46]. The most complete framework for achieving sustainable goals should be prioritized in sourcing sustainable biomaterials that provide a complete set of criteria to ensure that fuels are truly sustainable. Since dependence on first- and second-generation feedstocks continues, especially in biofuel production, easy availability of these feedstocks is extremely important in terms of sustainability and achieving emission targets in SAF consumption. According to IEA data presented in **Figure 2.8**, vegetable oil consumption for biofuel production is expected to increase by 46% to 54 million tons in the period of 2022–2027, and the proportion of vegetable oil production to meet the increasing biofuel demand is expected to increase from 17 to 23% [2.47].

In addition, the demand for these feedstocks will further increase as used cooking oil (UCO) and animal fats, which are other sources of feedstocks, are in even higher demand due to their lower GHG emission intensity and their meeting of EU feedstock requirements [2.48]. Under this scenario, the use of UCO and animal fats consumes almost 100% of the estimated supply during the forecast period. Even when wastes among other feedstocks are taken into account (such as palm oil mill waste, sawdust oil, and other agricultural waste oils), the demand still increases to approximately 65% of the global supply [2.45, 2.47]. Ensuring the sustainability of raw material supply is crucial, as the potential of biofuels to contribute to global decarbonization efforts could be weakened if supply is inadequate. To achieve net-zero emission targets, feedstock diversity must be ensured. SAF production generates much lower emissions than conventional jet fuel production because SAFs are made from renewable feedstocks rather than drilled fossil oil. SAFs are produced from the most readily available and recyclable first- and second-generation feedstocks, such as UCO and biomass [2.49]. The use of these feedstocks reduces emissions from SAFs by up to 85% on a lifecycle basis compared to conventional jet fuel [2.48, 2.49]. The current demand for the supply of raw materials for SAF production cannot be met. As the SAF industry scales, additional raw materials accompanied by different conversion technologies will need to be commercially distributed to increase the supply to the industry. To integrate feedstocks directly into fuel production and to constantly meet the need, SAF production can be based on a mixture of feedstocks (as shown in **Figure 2.9**) [2.50].

Figure 2.8 Biofuel feedstock requirement in the period of 2010–2027 [2.47].

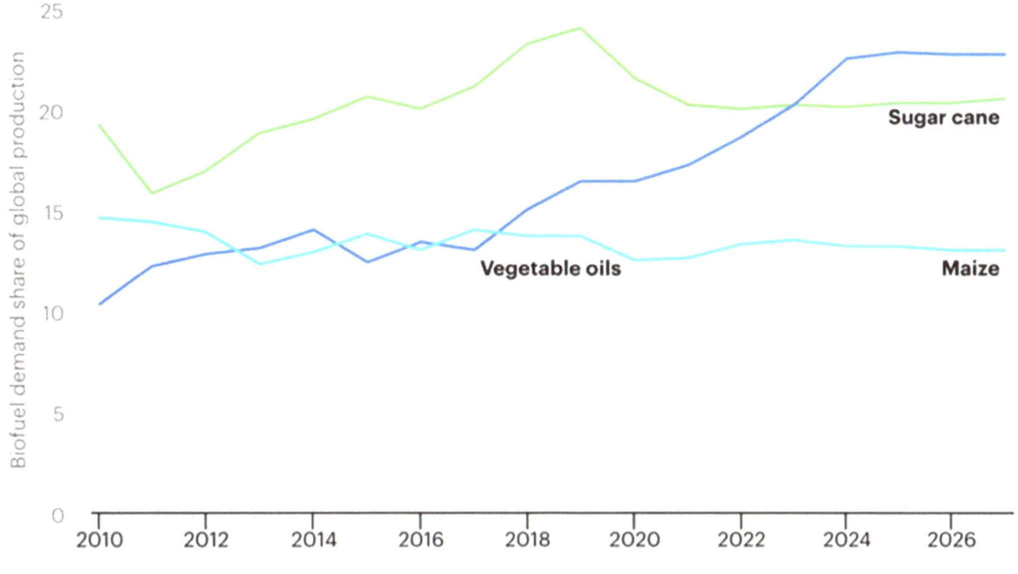

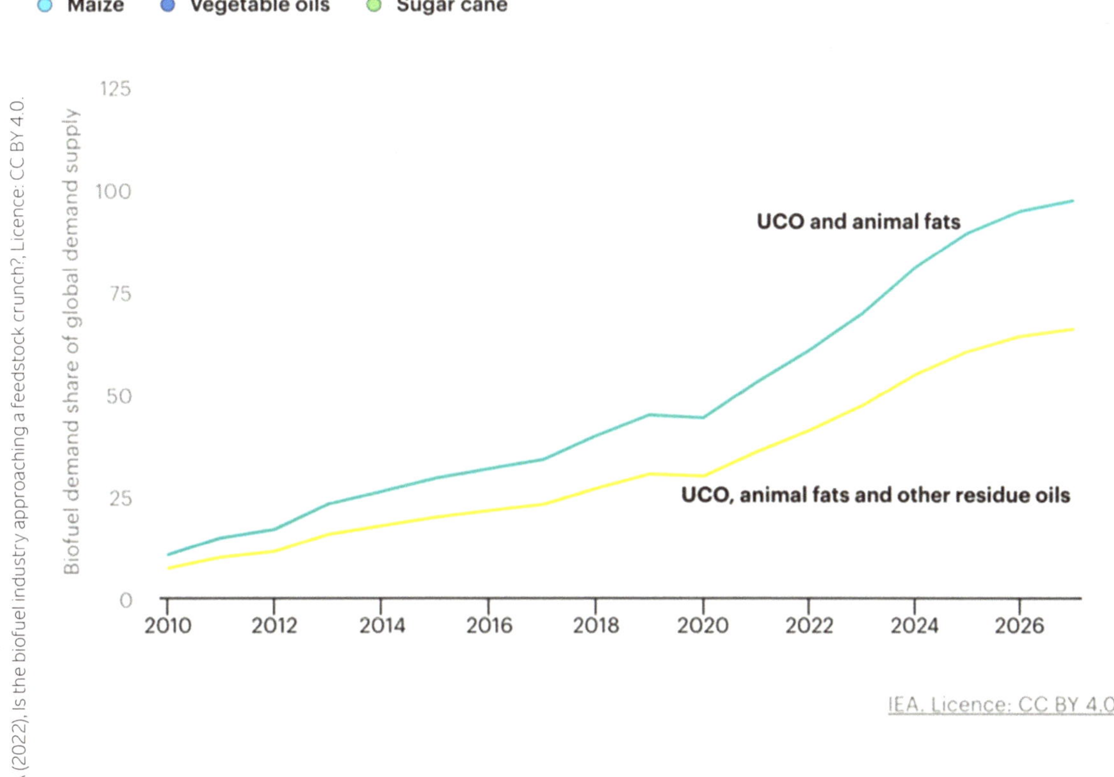

IEA (2022), Is the biofuel industry approaching a feedstock crunch?, Licence: CC BY 4.0.

Figure 2.9 Mixture of feedstocks for SAF production [2.50].

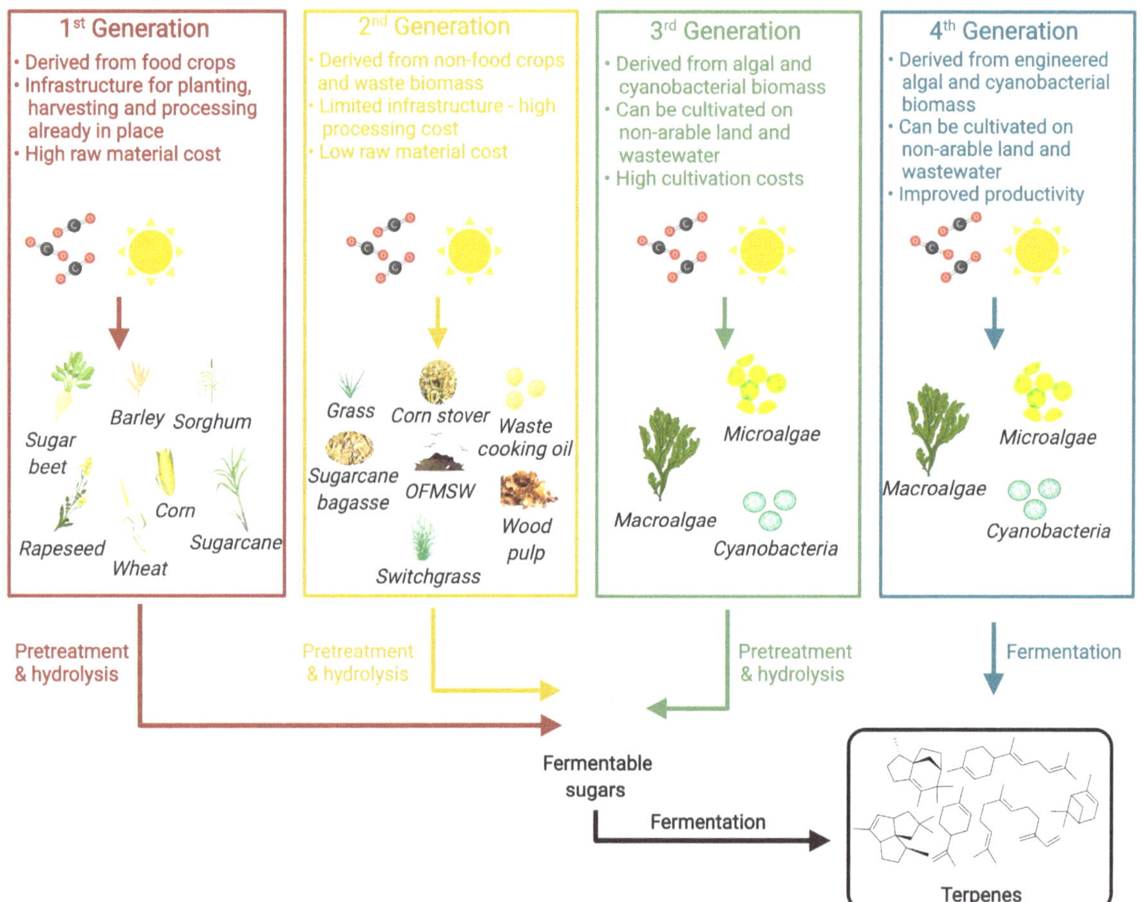

The biofuel standard, state LCFS, and tax credits are the factors increasing the demand for renewable diesel and biojet fuel in the US. In this context, in recent years, a large proportion of the need has been met by domestic production from first- and second-generation feedstock mixtures (such as soybean oil, rapeseed oil, corn oil, UCO, and animal fats) [2.51]. In Europe, the consumption of renewable diesel and biojet fuel is increasing. The increase in the overall demand is relatively small, but the EU is phasing out palm oil and imposing limits on other raw materials. This increases production from available waste, residues, and rapeseed oil [2.52]. To meet the increasing fuel needs of two major continents, studies continue to combine feedstocks to achieve the final product in a cheaper and more sustainable manner. At this point, sustainable feedstocks are the key to SAF production. Producing all generations of feedstocks by growing industrial and energy plants that are inedible and do not cause ILUC, along with the help of advanced SAF production technologies, will reduce CO_2 emissions and achieve sustainable development. A coordinated effort is needed to achieve the 2030 (near-term) and 2050 (long-term) net-zero emission targets, expand raw material availability and supply, reduce feedstock costs, increase raw material yields, improve sustainability, and provide economic opportunities among feedstock suppliers and producers [2.48]. In the near term, policy support for SAFs is needed to increase production and enable the widespread use of an expanded range of easily available and sustainable lipid-based feedstocks. In the conversion of feedstocks, there must be interaction with other areas of action on the roadmap, including SAF production technology innovation, creation of supply chains, and policy [2.51, 2.52].

As shown in **Figure 2.10**, this integration is achieved to meet the US Sustainable Aviation Fuel Grand Challenge Roadmap objectives, especially in terms of feedstock development [2.53]. Thus, innovations in feedstocks have facilitated innovations SAF-related feedstock types and sustainable raw feedstock supply systems. They also drive the production of SAF precursors such as ethanol and isobutanol by streamlining flows that enable supply chain optimization to reduce cost, technology uncertainty, and risk [2.54]. SAFs support feedstock supply and production, aiming for this roadmap to be the beginning of an evolving, collaborative, and necessarily dynamic process.

Figure 2.10 Sustainable Aviation Fuel Grand Challenge Roadmap of the US [2.53].

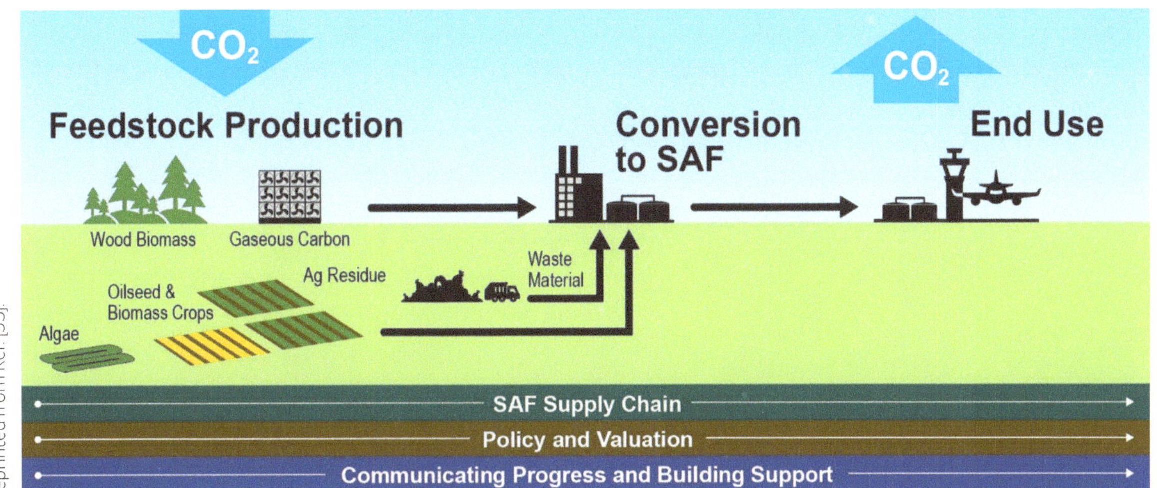

Conversion and Production Process

Collecting easily accessible feedstocks and choosing the appropriate conversion method for the raw material are basic steps in SAF production following the standards [2.55]. Since fuels that can be used in aircraft must have different properties from those used in marine and ground vehicles, the selected raw material is expected to meet the criteria for the final product. While aviation fuels must meet stringent requirements for flight characteristics such as flash point and freezing point, fuels in other sectors pass combustion waste through catalytic converters to reduce pollution [2.56]. When the consumption of commercial jet fuels on a flight basis and the ever-increasing fuel needs are evaluated together, the production of SAFs is of great importance [2.57]. It is accepted by the entire aviation sector management that reducing pollutant emissions, especially from fossil fuels, is a necessity to achieve net-zero emission targets. The long-term use of aviation fuels of the same quality at all airports is a fundamental factor in terms of sustainability. SAF options to fossil-based fuels are "drop-in" alternatives in that they have the potential to be a direct replacement for conventional jet fuel, requiring little or no changes to existing fuel and engine infrastructure or aircraft [2.58]. To ensure this, it is necessary to implement extremely strict quality control processes from raw materials to production in terms of both economic and environmental sustainability. The quality and affordability of feedstocks are of great importance to the developers of SAF production technologies. A conversion facility for SAF production must have a long lifespan and be sustainable in terms of the availability and acceptability of raw materials used in the process [2.55, 2.58]. If there is a problem in the supply of raw materials that will affect the process, the business situation of a commercial-scale facility consisting of many conversion technologies will quickly become problematic and the technologies developed in the process will face the danger of becoming obsolete [2.59]. However, in a long-term perspective, the development of an innovative conversion technology that simultaneously processes various similar and different types of raw materials will give a great boost to SAF production. The main advantage of such an approach is the provision of a consistent and stable supply of raw materials and durability and flexibility of such conversion facilities [2.60]. The implementation of such technologies will not depend on the location, the type of raw materials available, or even the season, which will help reduce financial risks for the producing parties. To meet the growing demand for net-zero SAFs, new conversion technologies and strategies are needed to process waste feedstocks, meet carbon reduction and cost targets, produce fuels that are compatible with existing engine technologies, and meet jet fuel specifications [2.59, 2.60]. Evaluating the entire process together and ensuring integrity in biofuel production, as shown in **Figure 2.11**, will provide the sustainability needed in the aviation sector [2.61].

Feedstocks are converted into SAFs through various conversion processes. As a result of this conversion, fuels such as bioalcohol, biodiesel, and biogas that can be used in ICEs and gas turbines are obtained (**Figure 2.12**) [2.62]. Today, these fuels are used by mixing them with petroleum-derived jet fuel in certain proportions within the standards [2.61, 2.62]. However, as the technology in the production phase improves and production costs decrease, SAFs may have the potential to become the main fuel of the aviation industry in the near future.

Figure 2.11 Biofuel feedstock production, certification, and use [2.61].

Figure 2.12 Biofuel products [2.62].

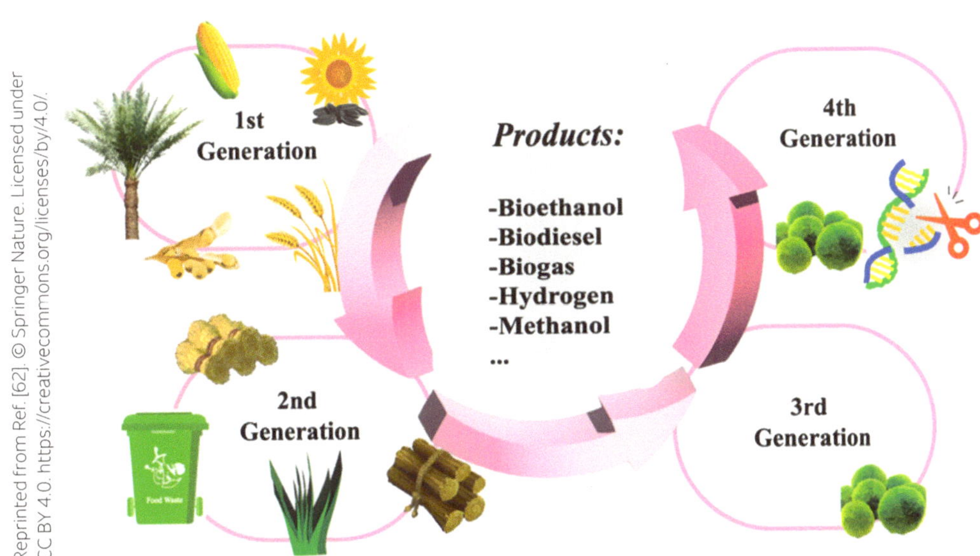

The fuels used in aircraft undergo strict testing to ensure flight safety and efficient engine operation. ASTM D1655 Standard Specification for Aviation Turbine Fuels defines properties, specifications, performance characteristics, test and sampling methods, etc., of petroleum-based aviation turbine fuels for civil use (Jet A and Jet A-1) [2.63]. The technical certification process for petroleum-based fuels is carried out according to the ASTM D1655 standard, so every SAF and fuel blended with conventional fuels must be certified under ASTM D1655 [2.64]. Drop-in fuels are combined with petroleum-based fuels either as a blend or, potentially in the future, as a 100% replacement. Therefore, the production of SAFs (drop-in), which will be used without any major modification in the airport fueling and engine fuel system infrastructures designed according to conventional fuels, must be carried out following a certain standard. ASTM International evaluates and approves SAF conversion processes and worldwide accreditations. SAFs must also meet many sustainability criteria to be eligible for use under ICAO CORSIA. As of July 2023, 11 conversion processes (see **Table 2.2**) for SAF production have been approved by ASTM, and 11 conversion processes are under evaluation under the current conditions [2.65].

Table 2.2 SAF conversion processes approved under the CORSIA framework [2.65].

ASTM reference	Conversion process	Abbreviation	Feedstocks	Maximum blend ratio (%)
ASTM D7566 Annex A1	Fischer–Tropsch hydroprocessed synthesized paraffinic kerosene	FT-SPK	Coal, natural gas, biomass	50
ASTM D7566 Annex A2	Synthesized paraffinic kerosene from hydroprocessed esters and fatty acids	HEFA-SPK	Vegetable oils, animal fats, used cooking oils	50
ASTM D7566 Annex A3	Synthesized isoparaffins from hydroprocessed fermented sugars	SIP	Biomass used in sugar production	10
ASTM D7566 Annex A4	Synthesized kerosene with aromatics derived by alkylation of light aromatics from non-petroleum sources	FT-SPK/A	Coal, natural gas, biomass	50
ASTM D7566 Annex A5	Alcohol to jet synthetic paraffinic kerosene	AtJ-SPK	Ethanol, isobutanol, isobutene from biomass	50
ASTM D7566 Annex A6	Catalytic hydrothermolysis jet fuel	CHJ	Vegetable oils, animal fats, used cooking oils	50
ASTM D7566 Annex A7	Synthesized paraffinic kerosene from hydrocarbon–hydroprocessed esters and fatty acids	HC-HEFA-SPK	Algae	10
ASTM D7566 Annex A8	Synthetic paraffinic ferosene with aromatics	AtJ-SPK/A	C2-C5 alcohols from biomass'	N/A
ASTM D1655 Annex A1	Co-hydroprocessing of esters and fatty acids in a conventional petroleum refinery	N/A	Vegetable oils, animal fats, used cooking oils from biomass processed with petroleum	5
ASTM D1655 Annex A1	Co-hydroprocessing of Fischer–Tropsch hydrocarbons in a conventional petroleum refinery	N/A	Fischer–Tropsch hydrocarbons coprocessed with petroleum	5
ASTM D1655 Annex A1	Coprocessing of HEFA	HEFA	N/A	10

The SAF component of the fuel must be certified at two separate stages before and after blending (**Figure 2.13**) under the ASTM D7566 and D1655 standards [2.66]. After validating a production methodology through a detailed and comprehensive certification process for SAFs, the conversion path is added as an annex to the above-mentioned standard [2.64, 2.67]. As shown in **Table 2.2**, there are currently eight main conversion pathways approved by the ASTM D7566 standard, allowing them to be used on a commercial scale, with the maximum blending percentage specified for each conversion technology. In addition, there is also the coprocessing of fat and oil feedstocks and petroleum-derived fuels in the existing refineries as three different conversion pathways to produce SAFs with up to 5% renewable content.

Figure 2.13 SAF blending certified process [2.66].

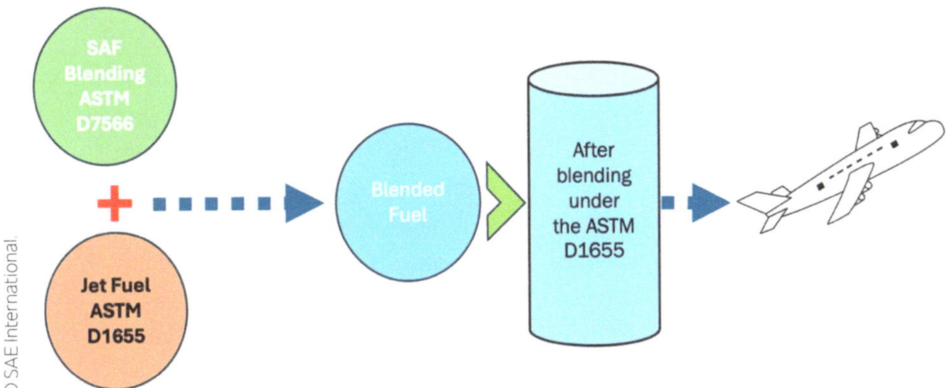

The selection of the targeted product market [ethanol, butanol, biofuel, fatty acid methyl ester (FAME), cheap fuels, etc.] in potential conversion technologies, including various routes through which there are alternatives for downstream processes, forms the basis for the selection of the conversion technology [2.68]. Different conversion pathways have different feedstock quality requirements, yield and production economies, and technical and operational challenges, as well as potentially a set of by-products and co-products, the marketing and sales of which may form an essential part of the business model. In addition to the main production routes, several transformation processes listed in **Table 2.3** are currently under evaluation by ASTM. In addition to ongoing production processes, efforts are also ongoing to enable the use of 100% SAF in aircraft and to increase the maximum blending ratio for coprocessing (from 5 to 30%) [2.69].

Table 2.3 SAF conversion processes under evaluation [2.69].

Conversion process under evaluation	Abbreviation	Lead developers
Synthesized aromatic kerosene	SAK	Virent
Integrated hydropyrolysis and hydroconversion	IH2	Shell
Single-reactor HEFA (drop-in liquid sustainable aviation and automotive fuel)	DILSAAF	Indian CSIR-IIP
Pyrolysis of nonrecyclable plastics	ReOIL	OMV
Coprocessing of pyrolysis oil from used tires	TPO	Philips 66
Methanol to jet	MTJ	ExxonMobil
Increase in fatty acid/ester coprocessing from 5 to 30%	—	—
HEFA with higher cycloparaffins	—	Revo
Biomass pyrolysis	—	Alder
Biomass/waste pyrolysis	—	Green Lizard
Cycloalkanes from ethanol	—	Vertimass

SAFs are currently certified to be blended with conventional jet fuel (up to 50%). In engines using old-type fuel systems, the mixture ratio is limited as a high mixture ratio will cause problems. However, there is no such concern with the new aircraft engine technology. Higher mixture limits are also likely to be approved in the near future as the use of synthetic aromatic compounds is approved and aircraft manufacturers are committed to making aircraft 100% SAF compliant by 2030 [2.70]. Researchers and practitioners in the aviation industry are constantly developing new methods to create SAFs at a lower cost with expanding technology and using different feedstocks [2.68, 2.70]. These new production pathways for the development and use of SAFs as the main means of decarbonization in aviation over the next few decades will support the increase in the SAF usage rate. Thus, as the capabilities of new production technologies continue to increase, more different types of SAFs will gain functionality, which will be able to position SAFs in the sector by finding a blend-based usage area as jet fuel [2.71].

Production Process

In contemporary practices of obtaining useful products from biomass, there are three main conversion processes to transform biomass into different products, namely thermochemical conversion, biological conversion, and chemical conversion [2.72]. The choice of conversion route is determined by the final product planned to be obtained.

Various processes are available to convert biomass into biofuel (**Figure 2.14**) such as thermochemical processes; gasification (syngas), pyrolysis (bio-oil, biochar, syngas), hydro-thermal carbonization (hydro char), hydrothermal liquefaction (io-oil), steam reforming (H_2), biochemical processes; anaerobic digestion (biogas), fermentation (ethanol), and chemical process; transesterification (biodiesel) [2.73]. The main factors in choosing these processes are the amount of raw materials and easy access to their source. For SAF production, a process that meets nine criteria at the optimum point, as shown in **Figure 2.15**, is put into practice [2.74]. SAF production routes are determined according to the biomass conversion method. All criteria are completely interconnected. What distinguishes these criteria from others is the validation of ASTM criteria required for the safe use of the end product of aviation fuels.

Figure 2.14 Biofuel production processes from biomass [2.73].

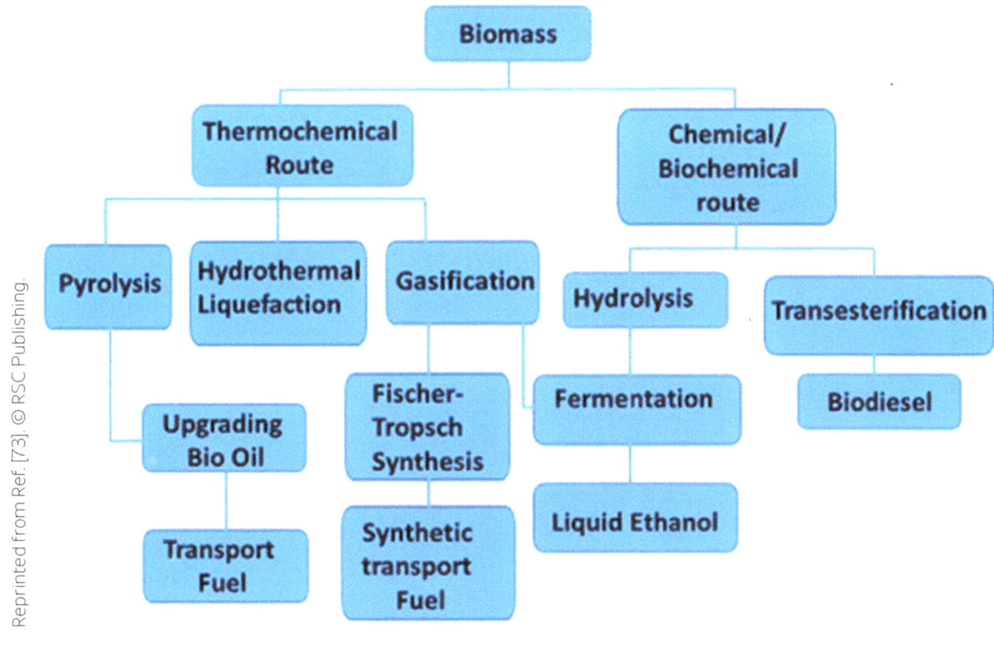

Figure 2.15 Production route criteria for SAF production [2.74].

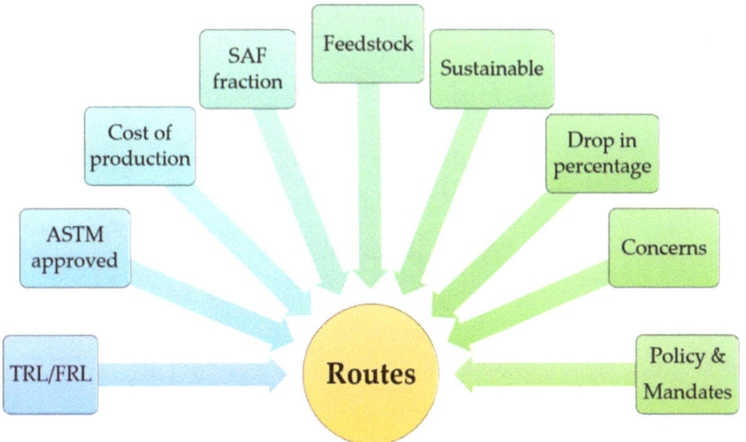

To produce fuels approved by ASTM, there are multiple conversion processes and blending limitations. These routes are categorized according to the technology, raw materials used to produce the fuel, and the cost of production. Each production method has different production efficiencies and costs [2.75].

Based on biomass conversion methods, thermo-chemical, biochemical, and chemical conversion pathways meeting the criteria that determine the production route for SAF production according to the type of feedstocks are specified as shown in **Figure 2.16** [2.76].

Figure 2.16 shows that, in the chemical conversion method, waste/used oils (UCO) and oils obtained from other biomass are used in the production process called oil conversion.

Figure 2.16 Main SAF production pathways [2.76].

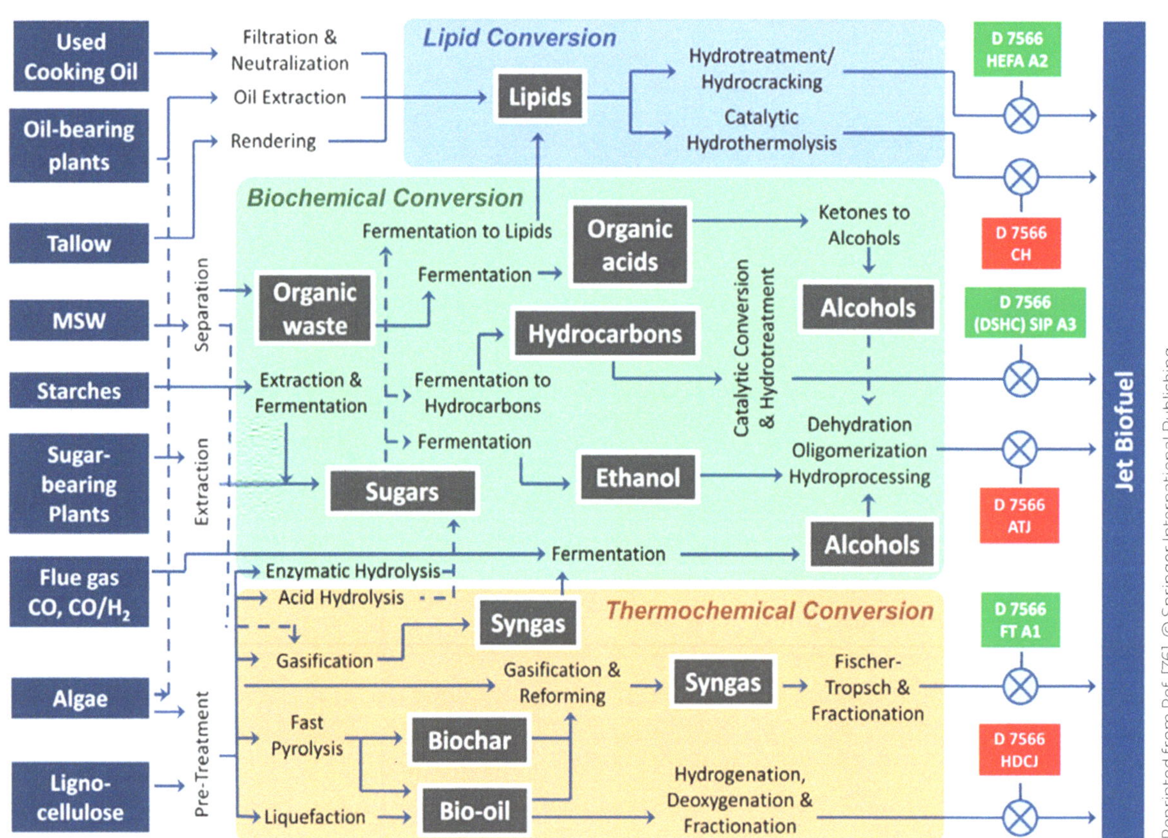

Fischer–Tropsch Synthetic Paraffinic Kerosene (FT-SPK, 2009)

The first Fisher–Tropsch production pathway, developed in 2009 as a thermochemical biomass conversion method, is a pioneering method in SAF production. These fuels are currently the best-known SAFs and have been tested and received ASTM certification for the large-scale production of aviation biofuels [2.77]. In addition, jet fuels are sometimes classified as kerosene (Jet A, Jet A-1, JP-5, and JP-8) or naphtha-type (Jet B and JP-4). The fuel produced must be referred to as synthetic paraffinic kerosene (SPK) after the approval process. Fuel with this abbreviation refers to any non-petroleum-based fuel designed to replace kerosene jet fuel [2.78].

FT-SPK Gas-to-Jet: Gasification of Biomass → FT → Hydroprocessing

During gasification, biomass is reacted with oxidants (most commonly CO_2, steam, or air) in a ratio such that partial oxidation occurs, producing CO and H_2-rich gas, also known as syngas. The CO–H_2 ratio is managed by a water–gas shift reaction, and subsequent CO_2 removal is made. Then, FT synthesis takes place [2.79]. These reactions occur in the presence of metal catalysts, typically at temperatures of 150–300°C (302–572°F) and high pressure. In 2009, with ASTM D7566 Annex A1 approval and FT-SPK definition, its use was allowed in mixing ratios up to 50% [2.77, 2.78, 2.79, 2.80]. FT-SPK developers and fuel-producing companies include Sasol, Shell, Syntroleum, Synfuels, Rentech, Solena, and Red Rock Biofuels. As shown in **Figure 2.17**, the process is completed as a result of gasification using coal, natural gas, MSW, agricultural wastes, forest wastes, wood, and energy crops in the production pathway mentioned above [2.78]. In general, all types of biomass are suitable for FT SAF production, but agricultural and forestry residues have the lowest net CO_2 emissions. Waste can also be used as a raw material to produce synthesis gas. However, lifecycle emissions from this process are relatively high with current technology due to the number of additional processing steps needed [2.81]. It is of great importance that the feedstocks to be used in this production pathway are sustainable. However, many commercially installed facilities use FT with fossil fuel feedstocks such as coal and natural gas, and the technology to produce liquid transportation fuels. The FT process is an extremely important production route in both coal liquefaction and gas to liquid (GtL) technology for the production of liquid HC fuels. In other words, it has gained a place in the literature as coal to liquid (CtL) and GtL. Unlike the CtL and GtL process, it will provide basic environmental gain for SAFs produced from MSW using FT technology [2.82, 2.83]. According to the World Bank, there is a potential of more than 2 billion tons of MSW per year worldwide. This amount is expected to increase to 3.4 billion tons by 2050. However, although access to MSW as a feedstock is widely available worldwide and is generally a lower-cost feedstock than other raw materials, in some regions the aviation sector competes with other sectors, including the energy industry, for access to MSW. Therefore, in SAF production, it would be of great benefit to convert synthesis gas produced from sources including MSW into long-chain HCs economically suitable for SAF production via FT. Thus, the development of an easy-to-use and cost-effective FT technology that can work on both large and small scales will increase the SAF usage rate [2.81, 2.82]. The liquid fuel produced through the FT pathway does not contain aromatic components. It is not appropriate to use aromatic-free fuel production methods in aircraft engines without mixing them with Jet A-1, because the aromatic content of the fuel prevents leaks in the engine fuel system.

Figure 2.17 FT synthesis process [2.78].

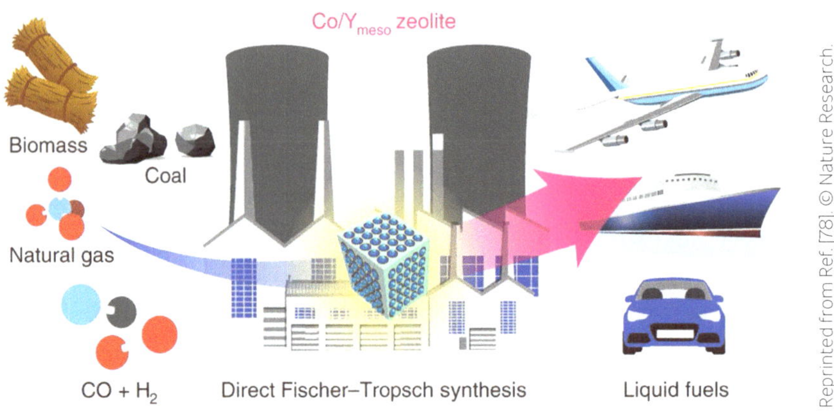

For this reason, as a result of the research to increase the aromatic content of FT-SPK fuel, the FT-SPK/A production process, approved by ASTM in 2015, was developed [2.80, 2.84]. FT has a Fuel Readiness Level (FRL) of 7. It has received approval in Appendix A4 of ASTM D7566 as another variation of the FT pathway.

FT-SPK/A, 2015

FT is the only route that currently has a validated process to directly incorporate aromatics into fuel [2.85].

FT-SPK/A Gas to Jet: Gasification of Biomass → FT → Hydroprocessing → Increased Aromatics Content

FT-SPK/A uses the same FT synthesis process but also involves the alkylation of light aromatics, particularly benzene. This results in the final product HC having an aromatic content, making it more compatible with existing engines in terms of preventing fuel leakage [2.86]. In other words, FT-SPK/A consists of blending FT-SPK with a small amount of aromatic HCs (around 20%) that match the properties of conventional jet fuel [2.85, 2.86]. At low temperatures, the wax content in FT-SPK can cause it to solidify and become too viscous, which can affect the fuel's flow properties and damage aircraft engines. Blending FT-SPK with aromatic HCs is also done to increase the cold-weather performance of the fuel [2.87]. From this perspective, FT-SPK/A is also a drop-in fuel. Theoretically, it has the potential to be used as a 100% trickle fuel without needing to be mixed with conventional jet fuel. However, although today it can only be used in blends up to 50%, it has the highest potential for increasing this value in the future and is likely to eventually be certified for 100% conventional fuel replacement [2.88]. FT-SPK/A technology developers and fuel-producing companies include Sasol, Shell, and Syntroleum. From an economic perspective, the advantage of using raw materials in a broad perspective increases the feasibility of establishing an FT plant because it allows the proportions of the products to be easily changed to maximize profits. This advantage in FT production cannot be fully utilized in SAF production since currently most of the progress in production is based on coal and natural gas as feedstocks [2.88, 2.89]. The FT production route using lignocellulosic biomass has the highest GHG emission savings potential, which could potentially be a solution in medium- and long-term plans of the aviation industry. In the future, this has the potential to allow higher blend percentages as it provides the full range of molecules found in existing jet fuels as opposed to just paraffins, thus providing better compatibility with existing jet engines [2.90]. However, due to limited technological maturity and high capital costs, further research and optimization of the production process are required before large-scale implementation [2.89, 2.90].

Hydroprocessed Esters and Fatty Acids to Synthetic Paraffinic Kerosene (HEFA-SPK, 2011)

One of the common production pathways for SAF production is HEFAs or hydroprocessed renewable jet (HRJ) or hydrogenated vegetable oil (HVO). Among SAF production technologies, HEFA SAF has an important place in terms of production potential due to lower capital costs and the availability of raw materials with high energy density [2.91]. HEFA fuels are currently the most commercially viable option in terms of the raw materials used and the technology in the production stages. In the HEFA production pathway (**Figure 2.18**), feedstocks containing triglycerides such as vegetable oil, animal fat, yellow or brown greases or waste oils, and algae oil are hydroprocessed to break down the long chain of fatty acids, followed by extraction and refinement, deoxygenation and hydrogenation, cracking and isomerization, and distillation [2.92]. Waste cooking oils as a source of feedstocks in HEFAs are currently the cheapest, most widely used oils with very high emissions reduction potential [2.91, 2.92].

Figure 2.18 HEFA process [2.92].

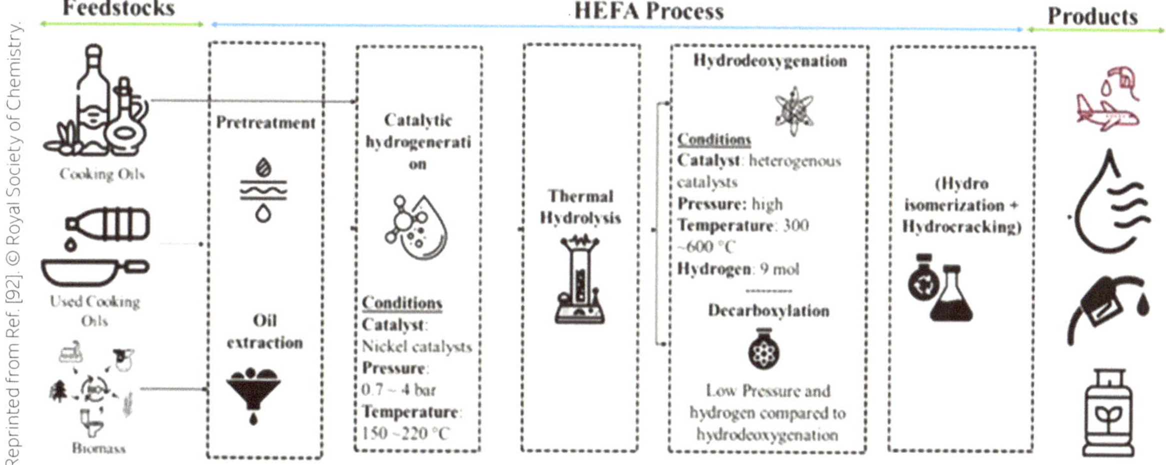

HEFA-SPK Oil-to-Jet: Deoxygenation of Oils and Fats → Hydroprocessing

The HEFA pathway was approved by ASTM in July 2011 with a 50% blending limit and has a FRL of 9 [2.93]. HEFA technology developers and fuel-producing companies include Neste Oil, Total, Honeywell UOP, Alt Air Fuels, Agrisoma, Biosciences, PetroChina, Sappire Energy, PEMEX, ASA, SG Biofuels, and Syntroleum. The HEFA pathway is also suitable for obtaining hydroprocessed renewable diesel, whereas jet fuel production requires additional fragmentation of longer-chain carbon molecules. This scenario brings about additional energy consumption and cost. Currently, biofuel is produced commercially through HEFAs. Most of this production focuses on producing biodiesel and/or SAF through HEFA, with capacity ranging from 0.1 million tons to approximately 100 million tons per year [2.94, 2.95]. Biodiesel, also known as HEFA, is produced by esterification from fatty acids, but it is considered inadequate as an aviation fuel because its energy density is significantly lower compared to conventional fuels and its freezing point is very high. This route provides promising SAF production in terms of time and economy and has drop-in potential. Another advantage of this process is that it can be integrated into oil refineries with just one additional step. This is the key reason why HEFA has led to more than 95% of all SAF flights and has emerged as the most commercially viable option for SAFs [2.96]. The fact that the raw material chosen for HEFA is a food source raises the question of whether it is food or fuel, as well as questions over its cost. This is the biggest obstacle to production that needs to be addressed because the cultivation of food-based raw materials can cause serious land-use change [2.95, 2.96]. Although HEFA-SPK is currently the only commercial route used on a large scale to produce SAFs, high-oil-content feedstocks need to be incorporated into the process to ensure the quality and sustainability of the fuel produced. To rapidly reach commercial, large-scale, sustainable raw materials, importance is given to the production of algae, *Camelina*, pennycress, tallow tree, and carinata among the alternative high-energy products that have been tried or already approved. The disadvantages of the fuel, such as low lubricity, are eliminated by blending it with conventional fuels. Due to its ease of production, the use of HEFAs as aviation fuel has been tested by many airline companies, giving it priority in passenger flights. Currently, the best-known process for SAF production is the HEFA pathway, which has undergone large-scale testing [2.97, 2.98].

Hydroprocessed Fermented Sugars to Synthetic Isoparaffins (HFS-SIP, 2014)

Unlike FT and HEFA fuels, HFS-SIP does not produce SPK. HFS-SIP is currently the most expensive among the existing production routes and has very high operating costs due to the necessary intermediate processing steps [2.99]. Before ASTM certification, this production route was called DSHC (direct sugar to HCs). Annex A3 of ASTM D7566 covers the certification of the use of SIP fuels and was approved in 2014. Although test flights using 20% mixtures have been successfully carried out, it is recommended to use mixtures not exceeding 10% in order to prevent fuel specifications from adversely affecting engine performance [2.100]. SAF is produced by the Amyris Company via the HFS route. HFS-SIP fuel is produced by fermenting sugars obtained from renewable feedstocks such as corn and sugarcane to produce high-carbon bioalcohols such as ethanol and butanol. Sugar, the main raw material of HFS-SIP, has a high availability performance [2.99, 2.100]. However, if sugarcane is used as a basis for sugar production, high land and water usage pose environmental risks [2.101]. Another disadvantage of this route is the high operating costs associated with the required processing steps. HFS-SIP is a multistep, multitechnology process that includes raw material preparation, fermentation, hydroprocessing, refining, and blending. As shown in **Figure 2.19**, during the hydroprocessing phase, various HCs are formed from bioalcohols [2.92]. This step is based on biochemical transformation that uses a biological platform to convert sugars into HCs through fermentation, specifically a C15 alkene with four double bonds called farnesene. In the final stage of the process, hydroprocessing converts it into the corresponding alkane farnesane, which is then included in the distillation process to obtain aviation fuel [2.102, 2.103].

Figure 2.19 HFS-SIP process [2.92].

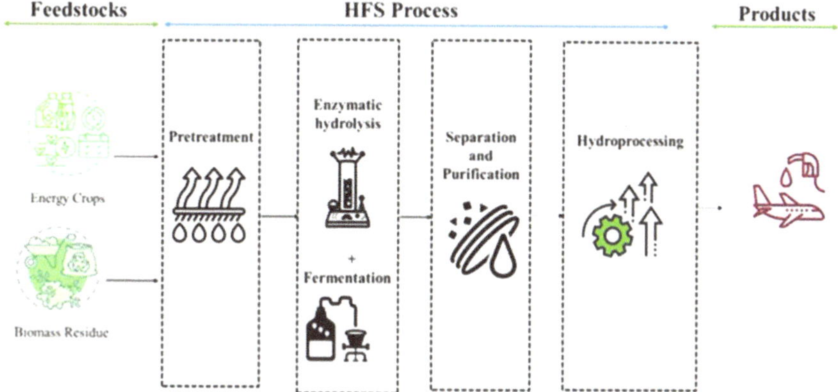

HFS-SIP Sugar-to-Jet: Hydrolysis of Biomass → Sugar Fermentation to Farnesene → Hydroprocessing → Fractionation

The complexity and low efficiency of converting lignocellulosic biomass-based sugars into fuels via DSHC result in high raw material costs and high energy consumption. These drawbacks have made DSHC the most expensive SAF production method [2.103].

Alcohol to Jet Synthetic Paraffinic Kerosene (AtJ-SPK, 2016)

The AtJ production route was approved as ASTM D7566 Annex A5 in 2016 under the designation AtJ-SPK [2.104]. AtJ-SPK SAF is produced by converting alcohols obtained from biomass, such as ethanol, into HCs that can be used to replace traditional jet fuel. Alcohols can be produced from cellulosic or starchy feedstock via fermentation or gasification reactions. It has a process in which biomass containing glucose and starch, such as sugarcane and corn grain, is converted into ethanol or other alcohols through fermentation, which can then be transferred or transported through pipes before being converted into AtJ-SPK [2.105].

AtJ-SPK Alcohol-to-Jet: Hydrolysis of Biomass → Sugar Fermentation to Alcohol → Dehydration → Oligomerization → Hydrogenation → Fractionation

As shown in **Figure 2.20**, the AtJ process involves the hydrolysis of wood fiber cellulosic biomass or glucose into intermediate alcohols (bioethanol, biobutanol, and fatty acid alcohols) and the production of biojet fuel through their dehydration and oligomerization [2.106]. Ethanol and isobutanol produced from lignocellulosic biomass are considered suitable feedstocks, but other potential feedstocks that have not yet received ASTM approval include methanol, isopropanol, and long-chain fatty alcohols.

ASTM approved it for isobutanol in April 2016 and for ethanol in June 2018, with a 30% blending limit. In the AtJ process based on bioethanol, the oxygen content of bioethanol is stabilized through dehydration, polymerization to access carbon atoms from existing petroleum-based aviation fuel, and hydrogenation to optimize physicochemical properties [2.105, 2.107]. Distillation is the final step, yielding an HC jet fuel, among other by-products, such as diesel. The maximum blend ratio currently allowed is up to 50%, with a FRL of 7. In the US, raw materials such as corn and sugarcane are already used commercially for fuel production. However, demand from sectors such as ground fuel and petrochemicals limits the use of raw materials in aviation [2.104, 2.106]. Crops used as feedstock for AtJ have high land- and water-use requirements, and these crops also lead to relatively high lifecycle carbon emissions due to indirect land-use considerations. A limited number of companies use the AtJ production route but do not produce SK commercially. Companies that produce SAFs using the AtJ production method are as follows: Honeywell UOP, Swedish Biofuels, Terrabon/MixAlco, Coskata, Solazyme, Kobalt, Gevo, LanzaTech, and Byogy Renewables. AtJ-SPK has the advantage of being produced from a wide range of biobased feedstocks, including waste materials such as agricultural and forestry residues. These raw materials are easy to grow and transport to the production line, but converting sugarcane into ethanol within 48 h after cutting is an important issue in terms of production potential [2.108].

Figure 2.20 AtJ process [2.106].

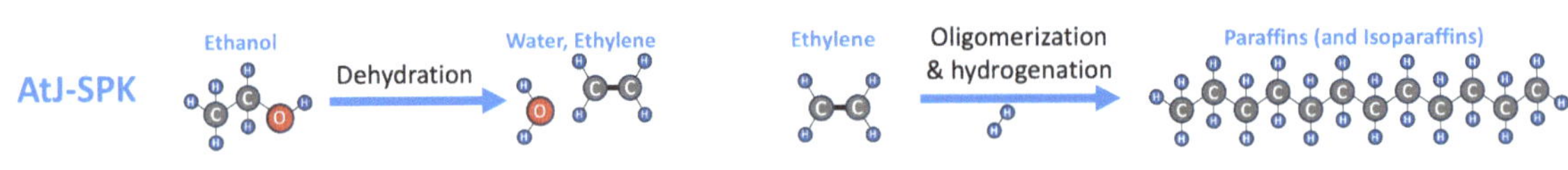

Courtesy of U.S. Department of Energy.

Whereas HEFA production is operationally simple, the AtJ process is multistep [2.107, 2.108]. To properly evaluate the economic feasibility of the AtJ conversion process, commercial production facility, upgrade process, and product distribution are the priority issues that require concentrated efforts for process upgrade. In order to achieve lower logistics costs, reduce carbon emissions due to transportation, and make better use of the infrastructure, ethanol plants should be located close to raw material production facilities and refineries, and integration should be ensured, as shown in **Figure 2.21** [2.109].

Figure 2.21 AtJ SAF production and drop-in integration [2.109].

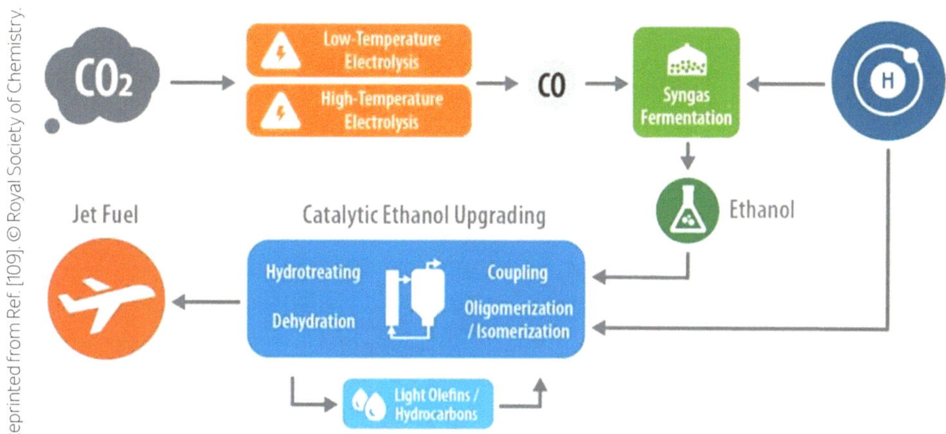

This integration is an important step in the profitability of SAFs. With raw material availability, manufacturers will focus on technological advancements that can reduce relatively high capital costs [2.105, 2.108]. Additionally, AtJ-SPK/A, certified as ASTM D7566 Annex A8 containing aromatic C2–C5 alcohols from biomass, is also being developed as a variant of the AtJ process [2.110]. Newly developed fermentation technologies may make the production of higher-carbon alcohols more cost-competitive than that of ethanol in the future. Thus, further research and development is needed to reduce the high production costs of AtJ and maximize its future benefits [2.108, 2.110].

Catalytic Hydrothermolysis Synthesized Kerosene (CH-SK, or CHJ; 2020)

Another fuel production pathway approved by ASTM is CHJ. It is also called hydrothermal liquation. CHJ-SK is typically produced by a process known as catalytic hydrothermolysis [2.111]. Feedstocks for the CH-SPK process can be a variety of triglyceride-based feedstocks such as soybean oil, *Jatropha* oil, *Camelina* oil, carinata oil, and tung oil [2.112]. Current CHJ feedstocks are typically energy crops or waste oils, with either high land and water use or limited availability. The process involves the heating of a mixture of water and biomass to high temperatures and pressures in the presence of a catalyst. CHJ proceeds via similar lipid intermediates to HEFAs, before reacting the lipids with water under extreme temperatures and pressures to produce a mixture of HCs. Catalytic hydrothermolysis is catalytically performed in the presence of water at a temperature ranging from 450 to 475°C and 21 MPa pressure [2.113]. The catalyst breaks down the complex organic molecules in the biomass into simpler aromatic HCs.

CH-SK Catalytic-to-Jet: Mixing of Biomass and Water → Hydrothermolysis (High Temperature and Pressure) → Catalytic Hydrocracking → Simpler HCs

After a subsequent hydrotreatment to saturate residual olefins and remove residual oxygenates, the liquid is distilled and fractionated into final HC products [2.114]. The resulting liquid is then refined and blended to create a fuel that can be used as a drop-in fuel. Biojet fuel produced from the catalytic hydrothermolysis pathway meets ASTM and military requirements thanks to its extraordinary cold fluidity properties, stability, and aromatic content [2.13]. Catalytic hydrothermolysis is one of the production pathways certified by ASTM (with annex A6) in 2020 for the latest commercial use. CHJ's maximum 50% blending rate of conventional jet fuel is recommended [2.115]. Ara, Euglena, and Aemetis/Chevron Lummus Global companies produce CHJ by pilot production. CHJ currently has a high cost of production due to its high capital and operating costs.

Hydroprocessed HCs, Esters, and Fatty Acids Synthetic Paraffinic Kerosene (HHC-SPK, or HC-HEFA-SPK; 2020)

The HC-HEFA-SPK or HHC-SPK process was approved by ASTM as Annex A7 of D7566 with fast-track approval in May 2020 and is allowed to be used in up to a 10% mixing ratio [2.115]. The process involves the production and use of a type of paraffinic kerosene synthesized from HC-HEFA. HC-HEFA works similarly to the HEFA process, starting with hydrodeoxygenation, followed by cracking and isomerization. The HC-HEFA pathway refers to the hydroprocessing of bioderived HCs (as distinct from fatty acids or fatty acid esters found in HEFA production) from oils found in a particular algae [2.116, 2.117]. This process converts it into more functional HCs by treating the feedstock with hydrogen to remove molecules that should not be present in the raw material, such as oxygen and other radicals [2.115, 2.117]. To create a synthetic jet fuel suitable for blending, HCs are cracked and isomerized, thus completing the process as shown in **Figure 2.22** [2.118]. Flow properties are also improved to increase the potential of the fuel obtained by using catalysts.

Figure 2.22 HHC production pathway [2.118].

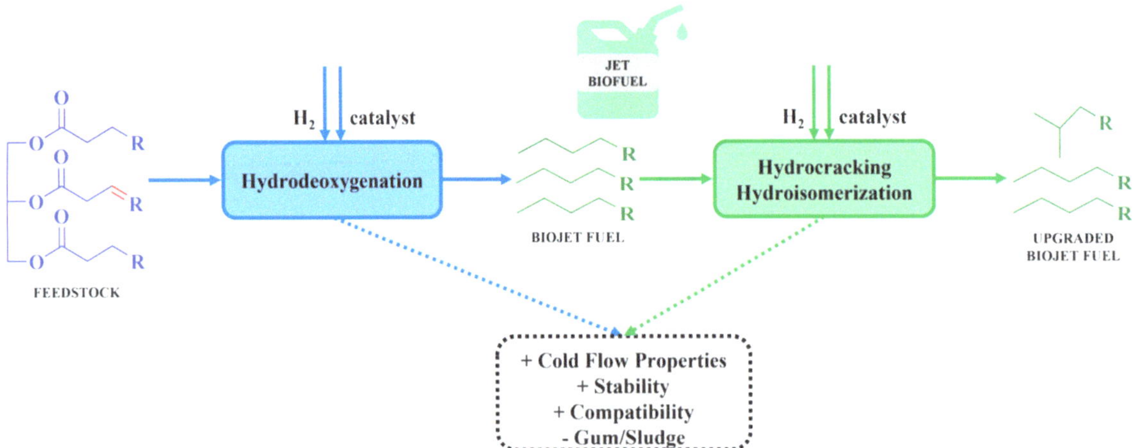

The HEFA production method using second-generation oilseed crops and waste oils could be an effective immediate solution with the potential to reduce significant GHG emissions [2.115, 2.117]. Moreover, processing of esters and fatty acids with biocatalysts may provide an interesting alternative for the production of HCs under mild reaction conditions. Microalgae oil as a feedstock could potentially offer much higher SAF efficiency and reductions in GHG emissions, but large-scale algae cultivation technology is currently not efficient enough [2.119]. However, further engineering is needed to identify optimized biocatalysts to obtain HCs in accordance with ASTM specifications, with high efficiency and product selectivity, and in an economically viable process [2.118, 2.119].

Co-processing SPK

The certification process specified (**Figure 2.23**) in ASTM D1655 Annex A1 includes the process of jointly incorporating resources that have the potential to be used as fuel raw materials in another process [2.120]. In this approach, three ways have been determined by ASTM:

- **FOG Co-processing:** Vegetable oils, animal fats, UCO from biomass processed with petroleum. ASTM approved 5% FOG co-processing with petroleum intermediates as a potential SAF pathway [2.121].
- **FT Co-processing:** FT HCs coprocessed with petroleum. ASTM approved 5% FT syncrude coprocessing with petroleum crude oil to produce SAF [2.122].
- **Co-processing of HEFA:** Feedstocks consisting of mono-, di-, and triglycerides, free fatty acids and fatty acid esters, including hydroprocessed synthesized kerosene up to 5% by volume in conventional petroleum refinery processes [2.121, 2.122].

As shown in **Figure 2.23**, the aim of coprocessing is to increase the proportion of renewable fuels in the final product, reduce the footprint of the process, and produce more SAF. However, no consensus has yet been reached on the sustainability of this approach, which is recommended to be investigated by ASTM [2.120, 2.121, 2.122]. Although the approach seems to encourage the use of SAF, it requires determining whether the feedstock will be suitable for coprocessing and the optimum point of both processes in the production stages. In order to use production quantity and resources effectively, coprocessing is expected to have an increasing effect on the quality of the resulting fuel. Although the potential for GHG reductions in this process exists, some companies argue that fuel obtained from coprocessing should not meet the SAF classification [2.123].

Figure 2.23 Co-processing biofuel production pathway [2.120].

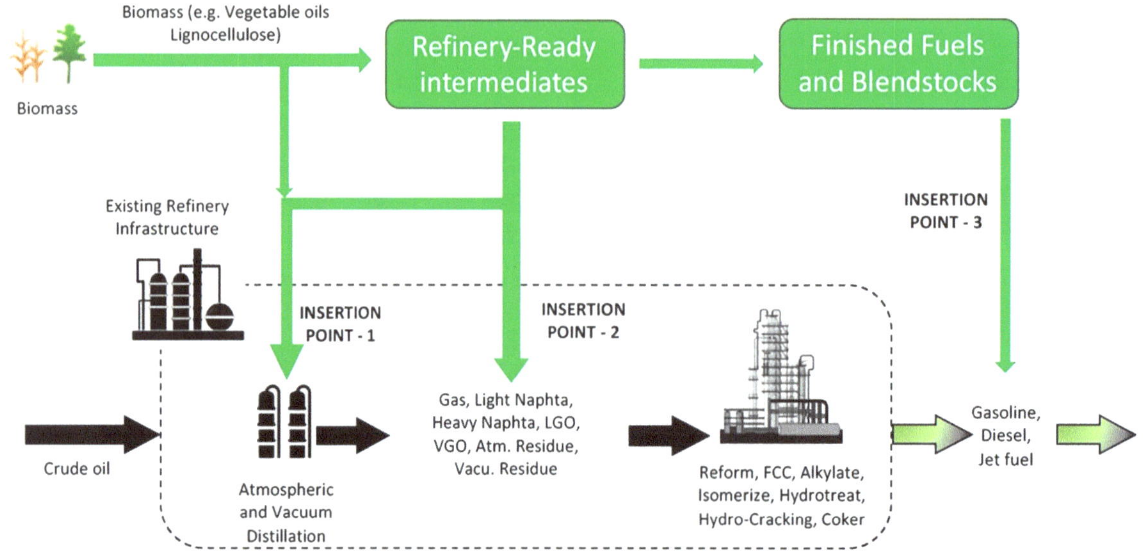

Advanced Production Technologies

The Commercial Aviation Alternative Fuels Initiative (CAAFI) encourages the development, evaluation, and commercialization of inexpensive alternative jet fuel options that offer equivalent levels of safety and compare favorably with petroleum-based jet fuel, while also offering environmental improvement and security of energy supply for aviation [2.124]. CAAFI FRL scale is based on NASA's Technology Readiness Level (TRL) scale and is intended to provide a classification to describe the progress of a conversion pathway toward commercialization [2.125]. For SAFs to replace traditional jet fuel, a number of key challenges must be overcome, including reducing costs, scaling the supply chain, and achieving 100% rather than proportionate use.

Various pathways have been developed to date to produce SAF from renewable biomass feedstocks. Depending on the biomass used, conversion routes have different production performances. As can be clearly seen in **Figure 2.24**, the TRL and FRL levels of ASTM-certified SAF production routes for commercial flights are the future projection of SAF available with current technology capabilities. Production pathways are available, but at different readiness levels. Based on the FRL level, the HEFA route currently has the most feasible technology and the lowest average cost. However, the HEFA route has a distinct disadvantage compared to other production routes, as it is the most cost-sensitive route in terms of raw material supply [2.127].

Figure 2.24 ASTM-certified pathways for commercial flights [2.126].

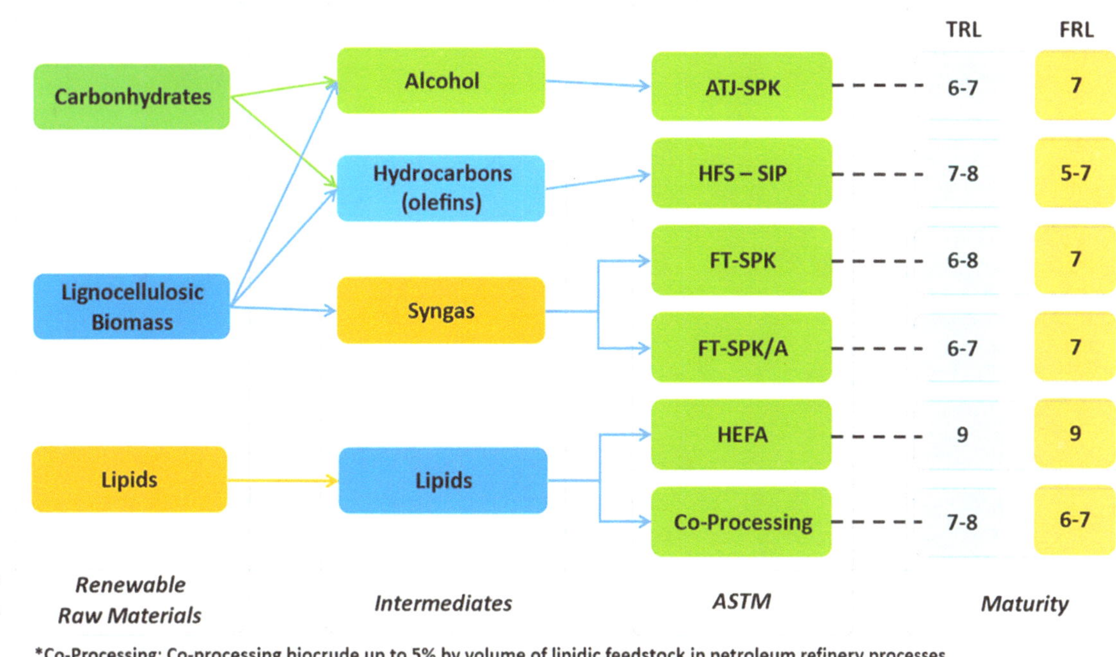

Fuel demand in the aviation industry is expected to increase in the near future, and how this demand should be met is considered an issue that needs to be resolved in order to comply with international efforts to reduce emissions in the sector. As shown in **Figure 2.25**, the Roland Berger SAF Sustainability Index evaluates the sustainability and cost of various SAF options separately, taking into account both current and next ten-year developments [2.128]. HEFA's dominance may not last long due to the rapid development of other transformation pathways and the rise of TRL. In particular, the FT production route has the potential to surpass HEFA as a production method open to commercial development [2.92, 2.129].

Figure 2.25 SAF production pathway sustainable index [2.128].

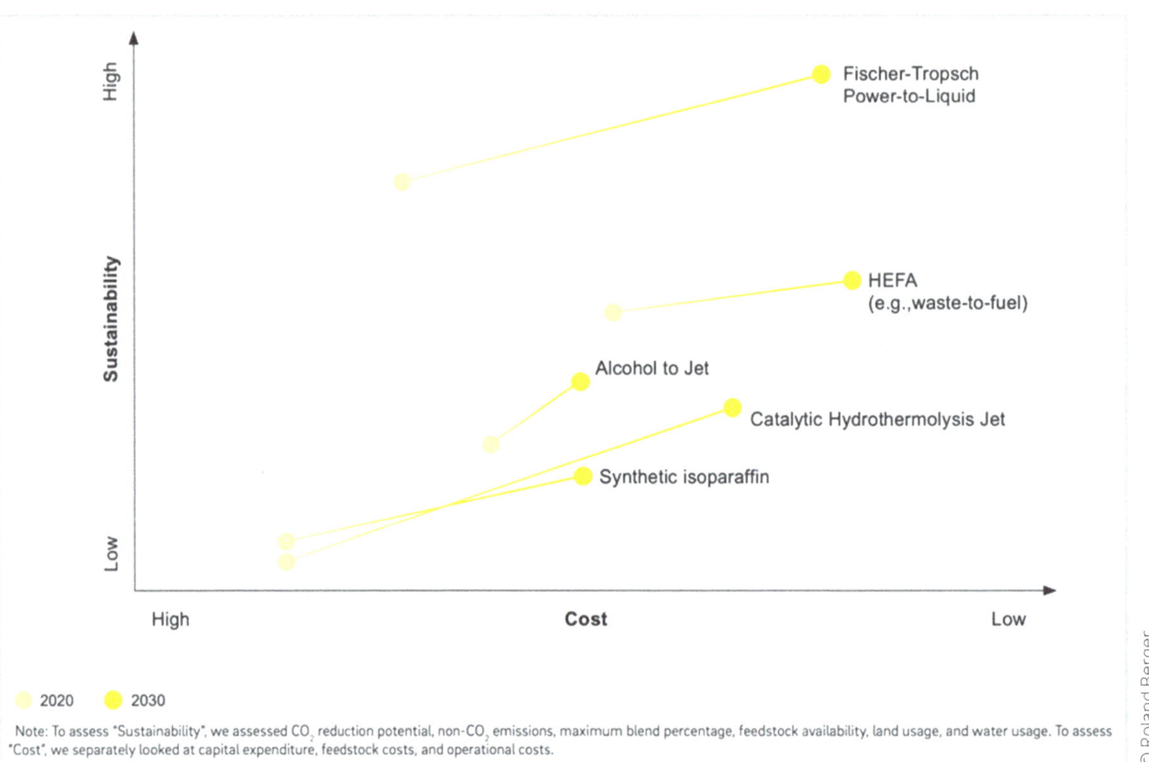

Although SAFs can also be produced from biomass, they have significant water and land requirements, so alternative feedstocks must be available that should not potentially compete with food crops [2.129]. Additionally, biofuels may cause higher GHG emissions than kerosene, primarily due to excessive land-use change for biomass. The aviation sector has ambitious targets for reducing its GHG emissions. GHG emission savings (in g CO_2 eq/MJ) measured in the evaluation made according to the types of feedstocks commonly used in some fuel production routes are shown in **Figure 2.26** [2.130].

Figure 2.26 GHG emission savings (in g CO_2 eq/MJ) by feedstocks types [2.130].

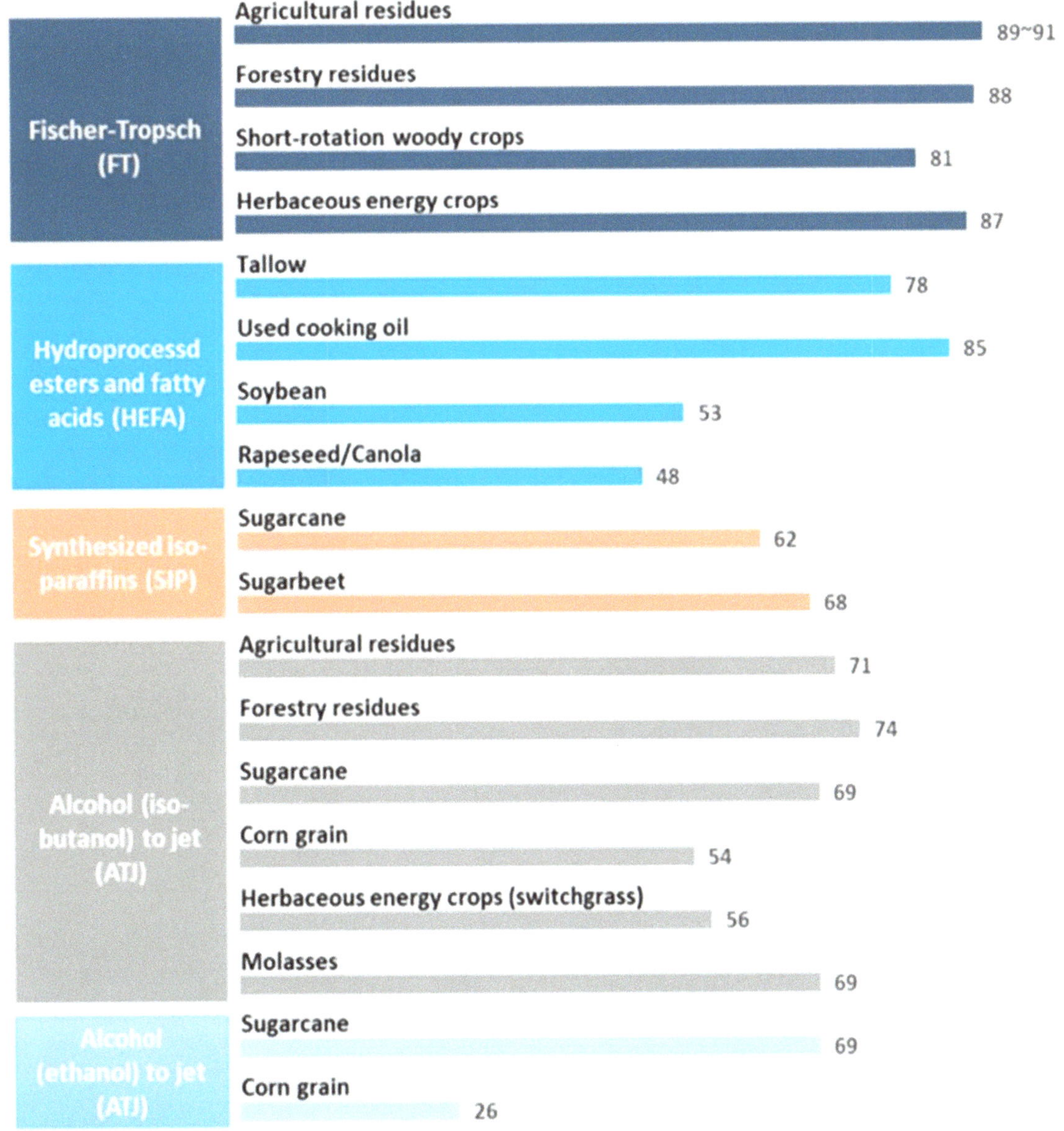

The reducing effect on GHG emissions is observed by selecting biomass suitable for the production route, ensuring its sustainability and inedible nature. The raw material source must be fed continuously to increase the production performance of the method that can be used commercially [2.58, 2.130]. When the use of raw materials and the chosen method are evaluated together, HEFA and FT production methods stand out commercially in terms of the future of SAF production and achievement of environmental targets [2.78, 2.91]. Given the huge resource demand, more advanced ways of producing SAFs are needed to meet the long-term demand of achieving a decarbonized aviation sector. Managing a company to commercialize SAFs involves many types of expenses, most of which fall into one of two categories: capital expenditures (CAPEX) and operating expenses (OPEX). These two values are particularly important for any organization in the oil and gas industry, where innovation, discovery, and global volatility impact many aspects of the industry. Considering the 2020 deployment, HEFA offers the most immediate solution [2.92]. It has the lowest CAPEX and OPEX among the three production routes studied due to its commercial maturity. In this context, the development of new SAF production routes continues rapidly, and continuous innovation is vital to reduce the cost of SAFs and increase their environmental benefits [2.71]. Although HEFA SAFs are currently the most suitable SAFs in reducing GHG emissions and production costs, studies are being carried out on more advanced production methods to ensure easier raw material supply in terms of sustainability [2.51, 2.131]. Many new routes that have entered or are being attempted for certification are generally aimed at improving the technology of existing routes in order to increase production efficiency, increase the maximum blending share, or facilitate production scales. Since the biomass used as raw material in known production routes has a relatively high oxygen content, a deoxygenation process is required to ultimately obtain HC liquid jet fuels with high volumetric energy density. In order to eliminate these processes in production, a hydrogen and CO_2 conversion process that can be integrated with technological methods developed in processes where production stages affect the cost and traditional methods, as shown in **Figure 2.27**, is being studied [2.132]. In addition, pilot-scale research is being carried out to determine the feasibility of using battery technology, which is accepted for use in land vehicles worldwide, in the aviation industry. Additionally, these advanced technologies must comply with current aviation standards and be used without major changes to both fuel systems and infrastructure facilities [2.59].

Such bioaviation fuels must have "drop-in" characteristics with strict specifications and compatibility with the combustion behavior of kerosene [2.60]. According to these mandatory criteria, the following emerging technologies, such as Power to Liquid (PtL), Sun to Liquid (StL), and Hydrogen-Electrification, are summarized, respectively.

Figure 2.27 SAFs that can be used in the aviation industry [2.132].

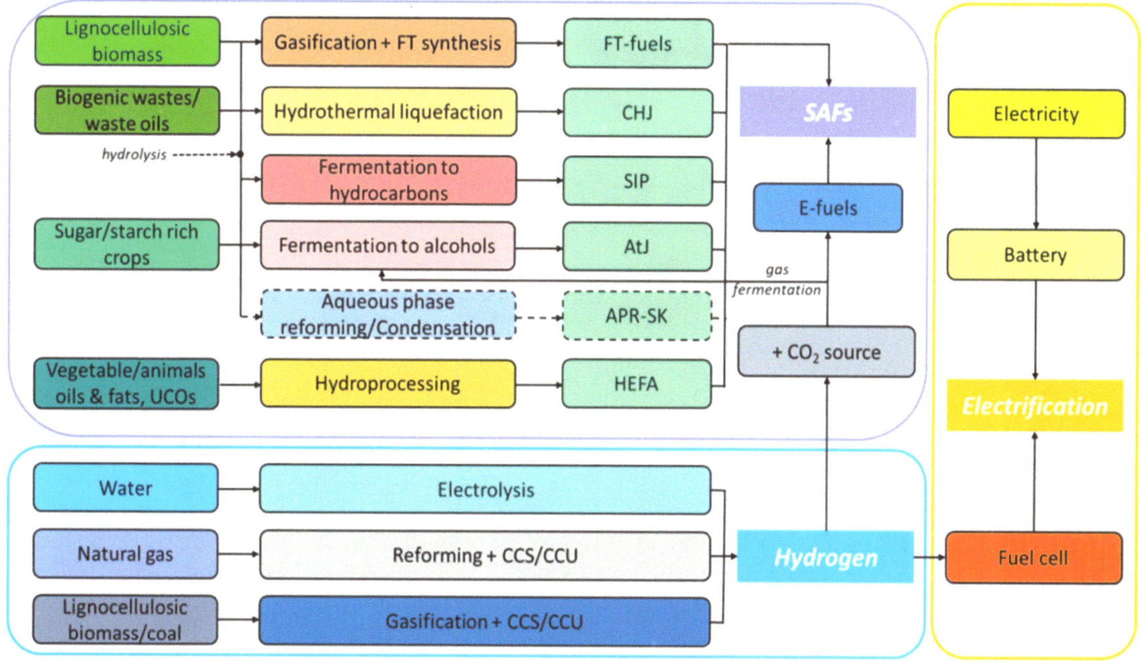

Power to Liquid (PtL)

In recent years, PtL, also called e-fuel, has been developed to meet the liquid fuel needs of ICEs as an alternative to battery technology with the help of advanced technologies [2.133]. e-fuels are a class of newly developed carbon-neutral replacement fuels produced by storing electrical energy from renewable sources in the chemical bonds of liquid or gaseous fuels [2.91]. There have also been initiatives to use e-fuels as SAFs. The aim of this production route is to meet the SAF needs in the aviation sector through PtL, an alternative fuel production route, without relying on biomass, arable land, and water. PtL obtained as a result of production is stated as a potential product that can be used instead of traditional jet fuel [2.90]. As shown in **Figure 2.28**, renewable electricity is the main energy source, and water and CO_2 are the main resources used in PtL production, which consists of three main steps [2.134]. First, renewable energy enables electrolyzers to produce green hydrogen. Methanol to gasoline (MtG) is a common process used in other industries that yields other middle distillate fuels (such as kerosene fuel). As shown in **Figure 2.28**, the MtK process is included in the synthetic fuel production line in the PtL process, similar to the FT process [2.135]. Improvement in the aromatic content of the fuel is achieved through this process. This produced fuel has the potential to contribute to achieving higher engine compatibility specifications and certification for higher blend ratios or potentially completely replacing conventional kerosene.

Figure 2.28 PtL SAF production pathway [2.135].

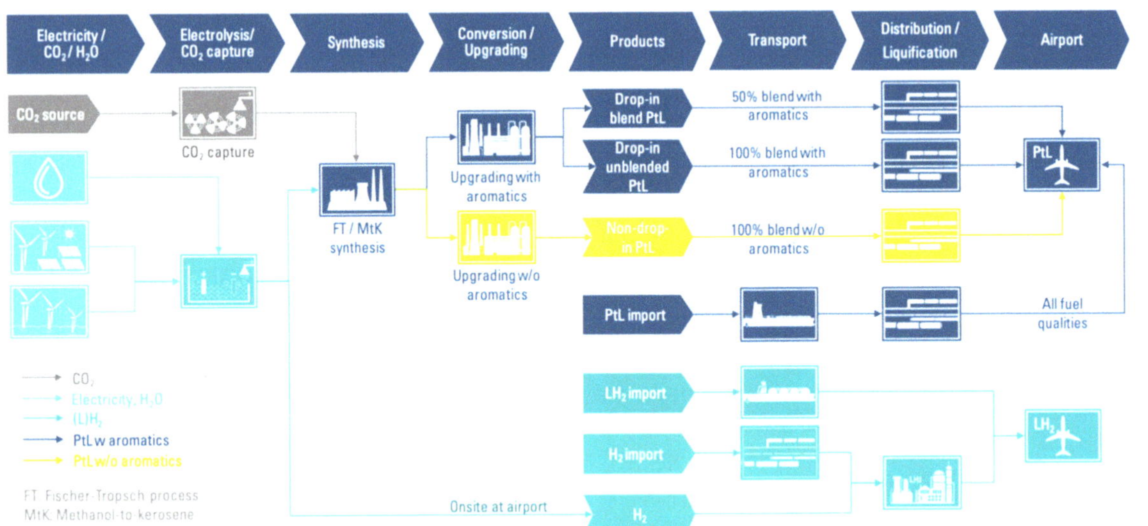

Climate-neutral CO_2 is then converted into carbon feedstock by carbon capture directly from the air. Carbon feedstocks are synthesized with green hydrogen through methods such as FT processes to produce liquid HCs. Finally, they are converted to produce a synthetic fuel similar to kerosene. PtL SAF produced from hybrid solar Photovoltaic (PV)-wind power plants via electrolysis and FT synthesis is a production route evaluated by ASTM [2.64, 2.65].

The most obvious factor in the development of this form of production depends on future production costs and, in particular, the cost of renewable electricity. The power part in the PtL production pathway determines the energy consumption in electricity production (the conversion of renewable electricity into liquid fuels), the separation of hydrogen, and the separation of carbon from CO_2 [2.133]. Capturing and storing CO_2 in the context of GHG mitigation is at the heart of PtL production. In effect, recapturing the CO_2 released during combustion and combining it with hydrogen closes the loop, with the CO_2 originally released being reused to create fuel. Therefore, with respect to the efficiency of different solutions regarding GHG emissions, PtL's GHG emissions can be reduced by up to 90% compared to fossil fuels. One of the biggest advantages of FT-integrated PtL is that it can be transported and distributed through the existing network of fossil fuel infrastructure, including pipelines and filling stations. Additionally, according to the ASTM-approved FT process and current jet fuel standards, it is possible to blend PtL with conventional kerosene (up to 50%) [2.63, 2.65]. The FT-PtL pathway requires synthesis gas production, and regular electrolysis must be applied to obtain H_2. Hydrogen production from the electrolysis of water is a fundamental step in the PtL process. Although it is currently far from being the most industrially used method to obtain H_2, it is a growing market and could become a future alternative for achieving carbon-free hydrogen production [2.133].

Compared with conventional aviation fuel, PtL production has a positive GHG balance and uses water and land more efficiently than the production of aviation alternative fuels made from biomass. Although PtL has this advantage, the most important difficulty with it on a commercial scale is that its production cost is currently higher compared to conventional aviation fuel. As their production increases, PtL fuels will be able to benefit from hydrogen infrastructure in both aviation and other industries, which will provide a significant economic advantage in the future. Thus, PtL fuels will be an important production route in the near future as aviation moves toward a more sustainable future and net-zero emission target.

Sun to Liquid (StL)

The sun to liquid (StL) process is a groundbreaking method to produce renewable synthetic fuel from CO_2, water, and sunlight. It consists of versatile technologies such as mirror field, solar receiver, thermochemical reactor, and thermal energy storage, as shown in **Figure 2.29** [2.136]. This production route promises a significant transformation toward the production of carbon-neutral synthetic fuels that release as much CO_2 as CO_2 extracted from the air. This process uses the entire solar spectrum and operates at high temperatures, providing a three-stage conversion suitable for fuel production from solar energy with high reaction rates and energy efficiency.

Figure 2.29 StL SAF production pathway [2.136].

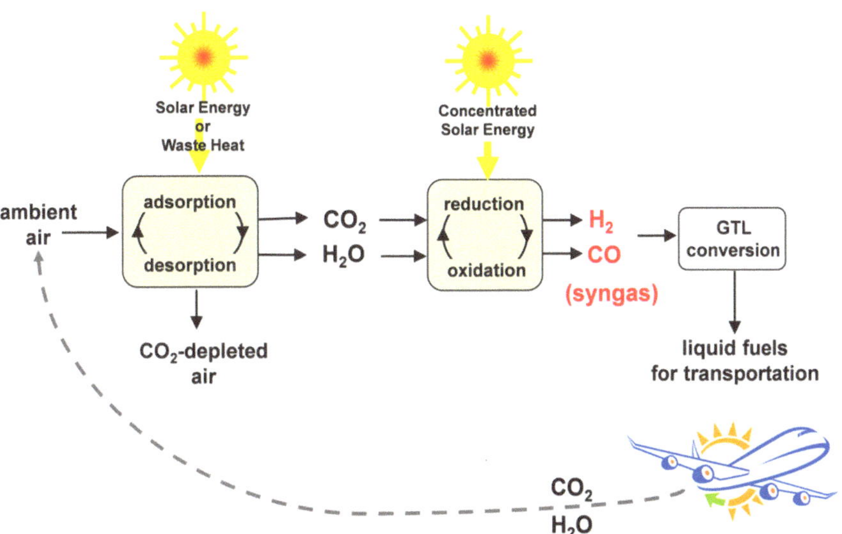

StL processes differ from PtL pathways because solar energy is used directly to synthesize liquid HC fuels. After the reaction, liquid fuel production follows the PtL pathways. The two-stage redox cycle ensures that syngas and O_2 are produced in separate steps. This prevents the formation of explosive mixtures while eliminating the need for high-temperature gas separation [2.137]. A high-temperature solar reactor splits CO_2 and H_2O taken directly from the air using concentrated solar radiation, producing syngas (GtL) that is eventually processed into liquid HCs such as methanol and kerosene. Fuels produced through StL production are compatible with existing fuel distribution, storage, and usage infrastructures and can be used as SAF by mixing kerosene [2.138]. Thanks to this technology that is likely to be developed, there will be a robust and sustainable transformation path to produce high-quality renewable liquid fuel from the inexhaustible potential of solar energy. Although it is theoretically promising like PtL technology, it is currently not ready for commercialization due to the initial installation cost and subsequent operating processes.

Hydrogen and Electrification

With the increasing demand for net-zero SAF, new conversion technologies are needed to process waste feedstocks and meet carbon reduction and cost targets [2.21]. In order to achieve emission reduction targets in the aviation industry, hydrogen-based battery technologies need to be brought to the agenda, especially among the more revolutionary measures for energy production [2.139]. SAFs are defined as liquid fuel alternatives and are expected to play an important role in hydrogen and electric propulsion as alternative fuels instead of liquid fuels in the decarbonization of aviation [2.13].

The technological location of the energy sources to be used in power-generating machines, especially in terms of reducing GHG emissions, is clearly shown in **Figure 2.30** [2.128]. For hydrogen- and electricity-based technologies to be used in aircraft, it is of great importance to create a detailed cycle both commercially and environmentally. The use of hydrogen in aviation power units has the potential to reduce GHG emissions and increase efficiency as long as hydrogen is produced from renewable energy sources [2.140]. On the other hand, biohydrogen can be produced from various biomass sources by thermal and biochemical methods. It can be used as liquid fuel in turbo engines or FCs. This process can bring hydrogen to a potential that can support SAFs [2.141].

Figure 2.30 CO_2 emissions according to alternative sources [2.128].

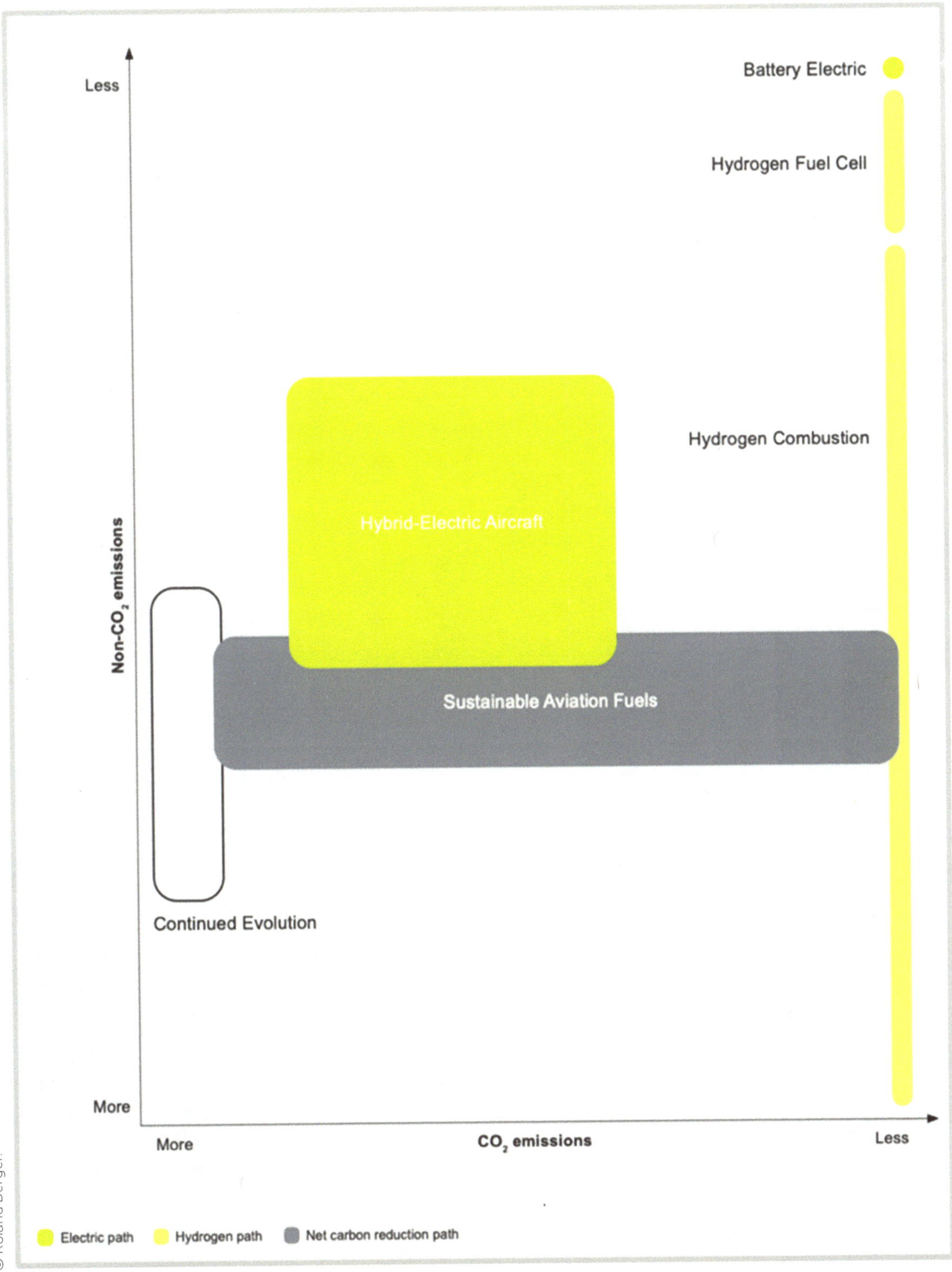

Liquid hydrogen is known to have a higher energy density than kerosene. In this regard, hydrogen can be used as fuel for aircraft or to power FCs. Although it is used instead of liquid fuel and the same combustion principles apply to the existing aircraft, large configurations or new designs in the engine will be inevitable due to the formation of waste gases and the differences in properties between kerosene and hydrogen [2.70]. On the other hand, hydrogen-electric engines, which have been newly developed and whose design and reliability have been tested on a pilot scale, have taken their place in the aviation literature as advanced propulsion products. In terms of the usage dynamics to meet the fuel needs in the aviation industry, it is clear that hydrogen-electric technology creates a great potential difference [2.142]. In parallel with similar technology development in land vehicles, hydrogen-electric technology with FC technology is theoretically becoming the best option for the long-term transition to clean aviation.

The hydrogen FC can provide continuous power production by generating electricity through the electrochemical reaction between hydrogen and oxygen. Among these cells, proton exchange membrane FCs (PEMFCs) and solid oxide fuel cells (SOFCs) have different operating principles, but both are theoretically determined to be suitable for use in aviation [2.143]. To ensure sustainability in the aviation industry, hydrogen must be available at the airport, independent of the existing fuel supply line. Producing, storing, transporting, and supplying hydrogen from renewable sources to aircraft also includes regulations that need to be addressed. In addition, a different infrastructure and distribution system should be developed [2.144].

Hydrogen-fueled aircraft stands as a promising and exciting development in the aviation industry, with the potential to significantly reduce carbon and other specific emissions and make air travel more sustainable [2.145]. However, many new technologies face serious technical challenges to replace the old ones. In particular, the energy density and the required amount of battery to obtain the same power increase the problem of weight. In this case, the problem of the aircraft struggling with the weight arises. While this situation can be tolerated in land vehicles, it cannot be solved in terms of extra safety and flight dynamics in aircraft, so there are challenges to battery-powered aircraft to have a promising future. Producing hydrogen fuel from renewable sources and its cost and availability remain challenges that need to be overcome. Both BE aircraft and FC-powered aircraft have not yet reached commercial potential due to flight range limitations [2.146]. The use of liquid hydrogen could extend flight distances, but the complexities of its storage and transportation on aircraft and on the ground need to be addressed extensively. Additionally, the transition to liquid hydrogen requires significant changes to the air transportation system, including new aircraft designs and adjusted airport infrastructure [2.147].

As a result, despite all these major issues, the latest developments in technology and partnerships between major players in the industry show that hydrogen-powered aircraft will play an important role in the future of aviation and achieving net-zero emission targets by 2035.

References

2.1. Organisation for Economic Co-operation and Development (OECD), "Carbon Management: Bioeconomy and Beyond," accessed October 21, 2024, https://www.oecd.org/en/publications/carbon-management-bioeconomy-and-beyond_b5ace135-en.html.

2.2. Khalifa, R., Alherbawi, M., Bicer, Y., and Al-Ansari, T., "Fueling Circularity: A Thorough Review of Circular Practices in the Aviation Sector with Sustainable Fuel Solutions," *Resources, Conservation & Recycling Advances* 23 (2024): 200223, doi:https://doi.org/10.1016/j.rcradv.2024.200223.

2.3. Zhang, X., Miao, W., Chen, H., Wang, J. et al., "Coupling of Crop Growth and Greenhouse Climate Models: Research and Progress in Intelligent Greenhouse Energy Saving," *Energy and Buildings* 329 (2025): 115252, doi:https://doi.org/10.1016/j.enbuild.2024.115252.

2.4. Amhamed, A.I., Al Assaf, A.H., Le Page, L.M., and Alrebei, O.F., "Alternative Sustainable Aviation Fuel and Energy (SAFE) - A Review with Selected Simulation Cases of Study," *Energy Reports* 11 (2024): 3317-3344, doi:https://doi.org/10.1016/j.egyr.2024.03.002.

2.5. Zhang, Y., Yanji Hao, Y.F., Feng, Y., Li, Y. et al., "GAN-MAML Strategy for Biomass Energy Production: Overcoming Small Dataset Limitations," *Applied Energy* 387 (2025): 125568, doi:https://doi.org/10.1016/j.apenergy.2025.125568.

2.6. The International Air Transport Association, "Current Airline Members," accessed October 29, 2024, https://www.iata.org/en/about/members/airline-list/.

2.7. The International Energy Agency, "Total Biofuel Production Growth by Region, Main Case, 2021-2027," accessed November 2, 2024, https://www.iea.org/data-and-statistics/charts/total-biofuel-production-growth-by-region-main-case-2021-2027.

2.8. Environmental and Energy Study Institute (EESI), "An Introduction to Sustainable Aviation Fuels," accessed November 4, 2024, https://www.eesi.org/articles/view/an-introduction-to-sustainable-aviation-fuels.

2.9. Dayil, J.L., Akande, O., Mahmoud, A.E.D., Kimera, R. et al., "Challenges and Opportunities in Machine Learning for Bioenergy Crop Yield Prediction: A Review," *Sustainable Energy Technologies and Assessments* 73 (2025): 104057, doi:https://doi.org/10.1016/j.seta.2024.104057.

2.10. Lv, X., Zhao, C., Yan, N., Ma, X. et al., "Sustainable Aviation Fuel (SAF) from Lignin: Pathways, Catalysts, and Challenges," *Bioresource Technology* 419 (2025): 132039, doi:https://doi.org/10.1016/j.biortech.2025.132039.

2.11. European Commission, "Renewable Energy Directive," accessed November 5, 2024, https://energy.ec.europa.eu/topics/renewable-energy/renewable-energy-directive-targets-and-rules/renewable-energy-directive_en.

2.12. Lte's Talk Science, "Introduction to Biofuels," accessed November 7, 2024, https://letstalkscience.ca/educational-resources/stem-explained/introduction-biofuels.

2.13. Shahriar, M.F. and Khanal, A., "The Current Techno-Economic, Environmental, Policy Status and Perspectives of Sustainable Aviation Fuel (SAF)," *Fuel* 325 (2022): 124905, doi:https://doi.org/10.1016/j.fuel.2022.124905.

2.14. Arushanyan, Y., Ekener, E., and Moberg, Å., "Sustainability Assessment Framework for Scenarios – SAFS," *Environmental Impact Assessment Review* 63 (2017): 23-34, doi:https://doi.org/10.1016/j.eiar.2016.11.001.

2.15. Cavelius, P., Engelhart-Straub, S., Mehlmer, N., Lercher, J. et al., "The Potential of Biofuels from First to Fourth Generation," *PLoS Biol* 21, no. 3 (2023): 3002063, doi:https://doi.org/10.1371/journal.pbio.3002063.

2.16. Yuchao, X., Zhang, Y., Deng, X., Lee, S.-Y. et al., "Bibliometric Analysis and Literature Review on Sustainable Aviation Fuel (SAF): Economic and Management Perspective," *Transport Policy* 162 (2025): 296-312, doi:https://doi.org/10.1016/j.tranpol.2024.11.014.

2.17. The International Civil Aviation Organization (ICAO), "SAF Feedstocks," accessed December 5, 2024, https://www.icao.int/environmental-protection/Pages/SAF_Feedstocks.aspx.

2.18. Jiang, C. and Liu, Y., "The Role of Sustainable Aviation Fuel in CORSIA: An Economic Analysis," *Energy Economics* 143 (2025): 108238, doi:https://doi.org/10.1016/j.eneco.2025.108238.

2.19. Cabulis, U. and Ivdre, A., "Recent Developments in the Sustainability of the Production of Polyurethane Foams from Polyols Based on the First- to the Fourth-Generation of Biomass Feedstock," *Current Opinion in Green and Sustainable Chemistry* 44 (2023): 100866, doi:https://doi.org/10.1016/j.cogsc.2023.100866.

2.20. Ghazali, M.F.S.M. and Mustafa, M., "Bioethanol as an Alternative Fuels: A Review on Production Strategies and Technique for Analysis," *Energy Conversion and Management: X* 26 (2025): 100933, doi:https://doi.org/10.1016/j.ecmx.2025.100933.

2.21. Xie, S., Li, Z., Luo, S., and Zhang, W., "Bioethanol to Jet Fuel: Current Status, Challenges, and Perspectives," *Renewable and Sustainable Energy Reviews* 192 (2024): 114240, doi:https://doi.org/10.1016/j.rser.2023.114240.

2.22. Zabed, H.M., Dar, M.A., Chowdhury, F.I., Xie, R. et al., "Bioethanol Production from Biomass: Introduction and Technologies," *Reference Module in Materials Science and Materials Engineering* (2025), doi:https://doi.org/10.1016/B978-0-443-29210-1.00020-0.

2.23. Gallejones, P., Pardo, G., Aizpurua, A., and del Prado, A., "Life Cycle Assessment of First-Generation Biofuels Using a Nitrogen Crop Model," *Science of the Total Environment* 505 (2015): 1191-1201, doi:https://doi.org/10.1016/j.scitotenv.2014.10.061.

2.24. Emmanouilidou, E., Lazaridou, A., Mitkidou, S., and Kokkinos, N.C., "A Comparative Study on Biodiesel Production from Edible and Non-Edible Biomasses," *Journal of Molecular Structure* 1306 (2024): 137870, doi:https://doi.org/10.1016/j.molstruc.2024.137870.

2.25. Yupanqui, K.R.G., Zeug, W., Thrän, D., and Bezama, A., "A Regionalized Social Life Cycle Assessment of a Prospective Value Chain of Second-Generation Biofuel Production," *Journal of Cleaner Production* 472 (2024): 143370, doi:https://doi.org/10.1016/j.jclepro.2024.143370.

2.26. Hombach, L.E., Cambero, C., Sowlati, T., and Walther, G., "Optimal Design of Supply Chains for Second Generation Biofuels Incorporating European Biofuel Regulations," *Journal of Cleaner Production* 133 (2016): 565-575, doi:https://doi.org/10.1016/j.jclepro.2016.05.107.

2.27. The International Energy Agency (IEA), "Sustainable Production of Second-Generation Biofuels," accessed December 13, 2024, https://www.iea.org/reports/sustainable-production-of-second-generation-biofuels.

2.28. Fei, Z., Chen, D., Hensen, E.J., Li, Y. et al., "Catalytic Valorization of Lignocellulose and Its Derived Feedstocks into Fuels and Chemicals," *Catalysis Today* 408 (2023): 1, doi:https://doi.org/10.1016/j.cattod.2022.10.021.

2.29. Microbiogen, "2nd Generation Biofuels," accessed December 23, 2024, https://microbiogen.com/article/2nd-generation-biofuels/.

2.30. Watson, M.J., Machado, P.G., da Silva, A.V., Saltar, Y. et al., "Sustainable Aviation Fuel Technologies, Costs, Emissions, Policies, and Markets: A Critical Review," *Journal of Cleaner Production* 449 (2024): 141472, doi:https://doi.org/10.1016/j.jclepro.2024.

2.31. Wandelt, S., Zhang, Y., and Sun, X., "Sustainable Aviation Fuels: A Meta-Review of Surveys and Key Challenges," *Journal of the Air Transport Research Society* 4 (2025): 100056, doi:https://doi.org/10.1016/j.jatrs.2024.100056.

2.32. Orfali, D.M., Meramo, S., and Sukumara, S., "Ranking Economic and Environmental Performance of Feedstocks Used in Bio-Based Production Systems," *Current Research in Biotechnology* 9 (2025): 100275, doi:https://doi.org/10.1016/j.crbiot.2025.100275.

2.33. Powar, R.S., Yadav, A.S., Ramakrishna, C.S., Patel, S. et al., "Algae: A Potential Feedstock for Third Generation Biofuel," *Materials Today: Proceedings* 63 (2022): A27-A33, doi:https://doi.org/10.1016/j.matpr.2022.07.161.

2.34. Varghese, G., Saeed, K., Xinming, L., and Rutt, K.J., "Measurements of the Tailpipe Emissions Characteristics of Algae Biodiesel Compared with the 1st and 2nd Generation Biodiesel Fuels," *Fuel* 388 (2025): 134315, doi:https://doi.org/10.1016/j.fuel.2025.134315.

2.35. Praveena, V., Martin, L.J., Matijošius, J., Aloui, F. et al., "A Systematic Review on Biofuel Production and Utilization from Algae and Waste Feedstocks – A Circular Economy Approach," *Renewable and Sustainable Energy Reviews* 192 (2024): 114178, doi:https://doi.org/10.1016/j.rser.2023.114178.

2.36. U.S. Environmental Protection Agency (EPA), "Overview of the Renewable Fuel Standard Program," accessed January 10, 2025, https://www.epa.gov/renewable-fuel-standard-program/overview-renewable-fuel-standard-program.

2.37. U.S. Department of Energy (DOE), "Advanced Algal Systems," accessed January 10, 2025, https://www.energy.gov/eere/bioenergy/advanced-algal-systems.

2.38. The Engineer, "EADS Aircraft Runs on Algae Biofuel," accessed January 10, 2025, https://www.theengineer.co.uk/content/news-analysis/eads-aircraft-runs-on-algae-biofuel/.

2.39. Ashokkumar, V., Chandramughi, V.P., Kumar, G., Ngamcharussrivichai, C. et al., "Advancements in Lignocellulosic Biomass: A Critical Appraisal of Fourth-Generation Biofuels and Value-Added Bioproduct," *Fuel* 365 (2024): 130751, doi:https://doi.org/10.1016/j.fuel.2023.130751.

2.40. Kazmi, A., Sultana, T., Ali, A., Nijabat, A. et al., "Innovations in Bioethanol Production: A Comprehensive Review of Feedstock Generations and Technology Advances," *Energy Strategy Reviews*

57 (2025): 101634, doi:https://doi.org/10.1016/j.esr.2024.101634.

2.41. Anto, S., Mukherjee, S.S., Muthappa, R., Mathimani, T. et al., "Algae as Green Energy Reserve: Technological Outlook on Biofuel Production," *Chemosphere* 242 (2020): 125079, doi:https://doi.org/10.1016/j.chemosphere.2019.125079.

2.42. Padder, S.A., Khan, R., and Rather, R.A., "Biofuel Generations: New Insights into Challenges and Opportunities in Their Microbe-Derived Industrial Production," *Biomass and Bioenergy* 185 (2024): 107220, doi:https://doi.org/10.1016/j.biombioe.2024.107220.

2.43. Eswaran, S., Subramaniam, S., Geleynse, S., Brandt, K. et al., "Techno-Economic Analysis of Catalytic Hydrothermolysis Pathway for Jet Fuel Production," *Renewable and Sustainable Energy Reviews* 151 (2021): 111516, doi:https://doi.org/10.1016/j.rser.2021.111516.

2.44. Bishop, G., Girón-Domínguez, C., Gaffey, J., Henchion, M. et al., "A Life Cycle Thinking-Based Environmental Risk Framework for Screening Sustainable Feedstocks in Early-Stage Bioeconomy Projects," *Resources, Environment and Sustainability* 20 (2025): 100201, doi:https://doi.org/10.1016/j.resenv.2025.100201.

2.45. Zahid, I., Nazir, M.H., Chiang, K., Christo, F. et al., "Current Outlook on Sustainable Feedstocks and Processes for Sustainable Aviation Fuel Production," *Current Opinion in Green and Sustainable Chemistry* 49 (2024): 100959, doi:https://doi.org/10.1016/j.cogsc.2024.100959.

2.46. Prussi, M., Lee, U., Wang, M., Malina, R. et al., "CORSIA: The First Internationally Adopted Approach to Calculate Life-Cycle GHG Emissions for Aviation Fuels," *Renewable and Sustainable Energy Reviews* 150 (2021): 111398, doi:https://doi.org/10.1016/j.rser.2021.111398.

2.47. The International Energy Agency (IEA), "Is the Biofuel Industry Approaching a Feedstock Crunch?" accessed December 11, 2024, https://www.iea.org/reports/is-the-biofuel-industry-approaching-a-feedstock-crunch.

2.48. zu Castell-Rüdenhausen, M., "Material Flow Analysis and Modelling of the Feedstock Potential for Recycling Polystyrene," *Circular Economy* 4, no. 1 (2025): 100127, doi:https://doi.org/10.1016/j.cec.2025.100127.

2.49. Racha, A., Kumar, L., Pai, S., Samanta, C. et al., "Highly Selective Hydrodeoxygenation Catalyst for Sustainable Aviation Fuel Production from Used Cooking Oil," *Catalysis Today* 442 (2024): 114895, doi:https://doi.org/10.1016/j.cattod.2024.114895.

2.50. Walls, L.E. and Rios-Solis, L., "Sustainable Production of Microbial Isoprenoid Derived Advanced Biojet Fuels Using Different Generation Feedstocks: A Review," *Front. Bioeng. Biotechnol.* 2020, no. 8 (2020): 599560, doi:https://doi.org/10.3389/fbioe.2020.599560.

2.51. Limb, B.J., Smith, J.P., Simske, S.J., and Quinn, J.C., "Estimating Geographic Origins of Corn and Soybean Biomass for Biofuel Production: A Detailed Dataset," *Data in Brief* 54 (2024): 110291, doi:https://doi.org/10.1016/j.dib.2024.110291.

2.52. Panoutsou, C., Germer, S., Karka, P., Papadokostantakis, S. et al., "Advanced Biofuels to Decarbonise European Transport by 2030: Markets, Challenges, and Policies that Impact Their Successful Market Uptake," *Energy Strategy Reviews* 34 (2021): 100633, doi:https://doi.org/10.1016/j.esr.2021.100633.

2.53. U.S. Department of Energy (DOE), "Synthetic Aviation Fuel Grand Challenge Roadmap: Flight Plan for Synthetic Aviation Fuel Report," accessed December 21, 2024, https://www.energy.gov/eere/bioenergy/articles/synthetic-aviation-fuel-grand-challenge-roadmap-flight-plan-synthetic.

2.54. Akter, H.A., Huang, Y.-K., and Dwivedi, P., "Developing a Supply Chain Model for Sustainable Aviation Fuel Using Logging Residues in Georgia, United States," *Forest Policy and Economics* 170 (2025): 103401, doi:https://doi.org/10.1016/j.forpol.2024.103401.

2.55. Mannion, L.A., Bell, A., Watson-Murphy, T., Kelly, M. et al., "A Physics Constrained Methodology for the Life Cycle Assessment of Sustainable Aviation Fuel Production," *Biomass and Bioenergy* 185 (2024): 107169, doi:https://doi.org/10.1016/j.biombioe.2024.107169.

2.56. Pearlson, M., Wollersheim, C., and Hileman, J., "A Technoeconomic Review of Hydroprocessed Renewable Esters and Fatty Acids for Jet Fuel Production," *Biofuels Bioproducts & Biorefining* 7 (2013): 89-96, doi:https://doi.org/10.1002/bbb.1378.

2.57. Dray, L., Schäfer, A.W., Grobler, C. et al., "Cost and Emissions Pathways towards Net-Zero Climate Impacts in Aviation," *Nat. Clim. Chang.* 12 (2022): 956-962, doi:https://doi.org/10.1038/s41558-022-01485-4.

2.58. Ruan, C., Yu, L., and Lu, X., "Towards Drop-In Sustainable Aviation Fuels in Aero Engine Combustors: Fuel Effects on Combustion Performance," *Progress in Aerospace Sciences* 153 (2025): 101054, doi:https://doi.org/10.1016/j.paerosci.2024.101054.

2.59. Blakey, S., Rye, L., and Wilson, C.W., "Aviation Gas Turbine Alternative Fuels: A Review," *Proceedings of the Combustion Institute* 33, no. 2 (2021): 2863-2885, doi:https://doi.org/10.1016/j.proci.2010.09.011.

2.60. Wei, C., Undavalli, V.K., Perkins, C., Heglas, K. et al., "Technical and Economic Assessment of Cryogenic Fuels for Future Aviation," *Progress in Aerospace Sciences* 153 (2025): 101053, doi:https://doi.org/10.1016/j.paerosci.2024.101053.

2.61. Stargate, "Sustainable Aviation Fuel Explained," accessed December 21, 2024, https://www.green-dealstargate.eu/news/sustainable-aviation-fuel/.

2.62. Wang, M., Ye, X., Bi, H. et al., "Microalgae Biofuels: Illuminating the Path to a Sustainable Future Amidst Challenges and Opportunities," *Biotechnol Biofuels* 17 (2024): 10, doi:https://doi.org/10.1186/s13068-024-02461-0.

2.63. American Society for Testing and Materials (ASTM), "ASTM D1655 Standard Specification for Aviation Turbine Fuels," accessed December 18, 2024, https://store.astm.org/d1655-22a.html.

2.64. Boehm, R.C., Faulhaber, C., Behnke, L., and Heyne, J., "The Effect of Theoretical SAF Composition on Calculated Engine and Aircraft Efficiency," *Fuel* 371 (2024): 132049, doi:https://doi.org/10.1016/j.fuel.2024.132049.

2.65. The International Civil Aviation Organization (ICAO), "Approved Conversion Processes," accessed December 18, 2024, https://www.icao.int/environmental-protection/GFAAF/Pages/Conversion-processes.aspx.

2.66. "Business Aviation Guide to the Use of Sustainable Alternative Jet Fuel (SAJF), SAJF Production Today," May 2018, accessed December 18, 2024, https://docs.wixstatic.com/ugd/7730ab_d501d6feb5eb45e-7b5a326111078412e.pdf.

2.67. Undavalli, V., Olatunde, O.B.G., Boylu, R., Wei, C. et al., "Recent Advancements in Sustainable Aviation Fuels," *Progress in Aerospace Sciences* 136 (2023): 100876, doi:https://doi.org/10.1016/j.paerosci.2022.

2.68. Gnanasekaran, L., Priya, A.K., Thanigaivel, S., and Hoang, T.K., "The Conversion of Biomass to Fuels via Cutting-Edge Technologies: Explorations from Natural Utilization Systems," *Fuel* 331 (2023): 125668, doi:https://doi.org/10.1016/j.fuel.2022.125668.

2.69. The International Civil Aviation Organization (ICAO), "A Conversion Processes under Evaluation," accessed December 18, 2024, https://www.icao.int/environmental-protection/GFAAF/Pages/Conversion-processes.aspx.

2.70. Air Transport Action Group (ATAG), "Sustainable Aviation Fuel," accessed December 21, 2024, https://atag.org/industry-topics/sustainable-aviation-fuel.

2.71. Aerospace Manufacturing and Design, "100% SAF," accessed November 12, 2024, https://www.aerospacemanufacturinganddesign.com/article/100-saf/.

2.72. Mahapatra, S., Kumar, D., Singh, B., and Sachan, P.K., "Biofuels and Their Sources of Production: A Review on Cleaner Sustainable Alternative Against Conventional Fuel, in the Framework of the Food and Energy Nexus," *Energy Nexus* 4 (2021): 100036, doi:https://doi.org/10.1016/j.nexus.2021.100036.

2.73. Bhutto, A.W., Qureshi, K., Abro, R., Harijan, K. et al., "Progress in the Production of Biomass-to-Liquid Biofuels to Decarbonize the Transport Sector – Prospects and Challenges," *RSC Adv.* 6 (2016): 32140-32170, doi:https://doi.org/10.1039/C5RA26459F.

2.74. Shehab, M., Moshammer, K., Franke, M., and Zondervan, E., "Analysis of the Potential of Meeting the EU's Sustainable Aviation Fuel Targets in 2030 and 2050," *Sustainability* 15, no. 12 (2023): 9266, doi:https://doi.org/10.3390/su15129266.

2.75. BP (Air BP), "How All Sustainable Aviation Fuel (SAF) Feedstock and Production Technologies Can Play a Role in Decarbonizing Aviation," accessed November 7, 2024, https://www.bp.com/en/global/air-bp/news-and-views/views.html.

2.76. Capaz, R.S., "Alternative Aviation Fuels in Brazil Environmental Performance and Economic Feasibility," Delft University of Technology, Brazil, Doctoral thesis, Delft University of Technology, 2021, https://doi.org/10.4233/uuid:a2740d8c-f08c-4fdd-96b8-54dd5d6fee01.

2.77. Bube, S., Bullerdiek, N., Voß, S., and Kaltschmitt, M., "Kerosene Production from Power-Based Syngas – A Technical Comparison of the Fischer-Tropsch and Methanol Pathway," *Fuel* 366 (2024): 131269, doi:https://doi.org/10.1016/j.fuel.2024.131269.

2.78. Li, J., He, Y., Tan, L. et al., "Integrated Tuneable Synthesis of Liquid Fuels via Fischer–Tropsch Technology," *Nat Catal* 1 (2018): 787-793, doi:https://doi.org/10.1038/s41929-018-0144-z.

2.79. Vervloet, D. et al., "Fischer–Tropsch Reaction–Diffusion in a Cobalt Catalyst Particle: Aspects of Activity and Selectivity for a Variable Chain Growth

Probability," *Catal. Sci. Technol.* 2 (2012): 1221-1233, doi:https://doi.org/10.1039/C2CY20060K.

2.80. Peters, M.A., Alves, C.T., and Onwudili, J.A., "A Review of Current and Emerging Production Technologies for Biomass-Derived Sustainable Aviation Fuels," *Energies* 16, no. 16 (2023): 6100, doi:https://doi.org/10.3390/en16166100.

2.81. Rafati, M., Wang, L., Dayton, D.C., Schimmel, K. et al., "Techno-Economic Analysis of Production of Fischer-Tropsch Liquids via Biomass Gasification: The Effects of Fischer-Tropsch Catalysts and Natural Gas Co-Feeding," *Energy Conversion and Management* 133 (2017): 153-166, doi:https://doi.org/10.1016/j.enconman.2016.11.051.

2.82. Kargbo, H., Harris, J.S., and Phan, A.N., "'Drop-In' Fuel Production from Biomass: Critical Review on Techno-Economic Feasibility and Sustainability," *Renewable and Sustainable Energy Reviews* 135 (2021): 110168, doi:https://doi.org/10.1016/j.rser.2020.110168.

2.83. SKYNRG, "Technology Basics," accessed November 21, 2024, https://skynrg.com/sustainable-aviation-fuel/technology-basics/.

2.84. Rumizen, M.A., "Qualification of Alternative Jet Fuels," *Front. Energy Res.* 9 (2021): 760713, doi:https://doi.org/10.3389/fenrg.2021.760713.

2.85. Advances Biofuels USA, "Fischer-Tropsch Synthetic Paraffinic Kerosene with Aromatics (FT-SPK/A)," accessed November 6, 2024, https://advancedbiofuelsusa.info/tag/fischer-tropsch-synthetic-paraffinic-kerosene-with-aromatics-ft-spk-a.

2.86. Liu, G., Yan, B., and Chen, G., "Technical Review on Jet Fuel Production," *Renewable and Sustainable Energy Reviews* 25 (2013): 59-70, doi:https://doi.org/10.1016/j.rser.2013.03.025.

2.87. Gupta, K.K., Rehman, A., and Sarviya, R.M., "Bio-fuels for the Gas Turbine: A Review," *Renewable and Sustainable Energy Reviews* 14, no. 9 (2010): 2946-2955, doi:https://doi.org/10.1016/j.rser.2010.07.025.

2.88. Jürgens, S., Oßwald, P., Selinsek, M., Piermartini, P. et al., "Assessment of Combustion Properties of Non-Hydroprocessed Fischer-Tropsch Fuels for Aviation," *Fuel Processing Technology* 193 (2019): 232-243, doi:https://doi.org/10.1016/j.fuproc.2019.05.015.

2.89. Heyne, J., Rauch, B., Le Clercq, P., and Colket, M., "Sustainable Aviation Fuel Prescreening Tools and Procedures," *Fuel* 290 (2021): 120004, doi:https://doi.org/10.1016/j.fuel.2020.120004.

2.90. Cabrera, E. and de Sousa, J.M.M., "Use of Sustainable Fuels in Aviation—A Review," *Energies* 15, no. 7 (2022): 2440, doi:https://doi.org/10.3390/en15072440.

2.91. Zheng, X., Shi, W., Wang, M., Zhong, S. et al., "Performance and Combustion Characteristics of Heavy-Fuel Aircraft Piston Engines at High Altitudes: Comparison between Conventional Fuels and HEFA Sustainable Aviation Fuel," *Sustainable Energy Technologies and Assessments* 75 (2025): 104210, doi:https://doi.org/10.1016/j.seta.2025.104210.

2.92. Usman, M., Cheng, S., Boonyubola, S., and Cross, J.S., "The Future of Aviation Soars with HTL-Based SAFs: Exploring Potential and Overcoming Challenges Using Organic Wet Feedstocks," *Sustainable Energy Fuels* 7 (2023): 4066-4087, doi:https://doi.org/10.1039/D3SE00427A.

2.93. Yang, F. and Yao, Y., "Sustainable Aviation Fuel Pathways: Emissions, Costs and Uncertainty," *Resources, Conservation and Recycling* 215 (2025): 108124, doi:https://doi.org/10.1016/j.resconrec.2025.108124.

2.94. Grimme, W., "The Introduction of Sustainable Aviation Fuels—A Discussion of Challenges, Options and Alternatives," *Aerospace* 10, no. 3 (2023): 218, doi:https://doi.org/10.3390/aerospace10030218.

2.95. Chu, P.L., Vanderghem, C., MacLean, H.L., and Saville, B.A., "Process Modeling of Hydrodeoxygenation to Produce Renewable Jet Fuel and other Hydrocarbon Fuels," *Fuel* 196 (2017): 298-305, doi:https://doi.org/10.1016/j.fuel.2017.01.097.

2.96. Han, J., Elgowainy, A., Cai, H., and Wang, M.Q., "Life-Cycle Analysis of Bio-Based Aviation Fuels," *Bioresource Technology* 150 (2013): 447-456, doi:https://doi.org/10.1016/j.biortech.2013.07.153.

2.97. Główka, M., Wójcik, J., Boberski, P., Białecki, T. et al., "Sustainable Aviation Fuel – Comprehensive Study on Highly Selective Isomerization Route towards HEFA Based Bioadditives," *Renewable Energy* 220 (2024): 119696, doi:https://doi.org/10.1016/j.renene.2023.119696.

2.98. Mäki-Arvela, P., Martínez-Klimov, M., and Murzin, D.Y., "Hydroconversion of Fatty Acids and Vegetable Oils for Production of Jet Fuels," *Fuel* 306 (2021): 121673, doi:https://doi.org/10.1016/j.fuel.2021.121673.

2.99. Eller, Z., Varga, Z., and Hancsók, J., "Advanced Production Process of Jet Fuel Components from Technical Grade Coconut Oil with Special Hydrocracking," *Fuel* 182 (2016): 713-720, doi:https://doi.org/10.1016/j.fuel.2016.06.055.

2.100. Braun, M., Grimme, W., and Oesingmann, K., "Pathway to Net Zero: Reviewing Sustainable Aviation Fuels, Environmental Impacts and Pricing," *Journal of Air Transport Management* 117 (2024): 102580, doi:https://doi.org/10.1016/j.jairtraman.2024.102580.

2.101. Farooq, D., Thompson, I., and Ng, K.S., "Exploring the Feasibility of Producing Sustainable Aviation Fuel in the UK Using Hydrothermal Liquefaction Technology: A Comprehensive Techno-Economic and Environmental Assessment," *Cleaner Engineering and Technology* 1 (2020): 100010, doi:https://doi.org/10.1016/j.clet.2020.100010.

2.102. Zhu, Y., Biddy, M.J., Jones, S.B., Elliott, D.C. et al., "Techno-Economic Analysis of Liquid Fuel Production from Woody Biomass via Hydrothermal Liquefaction (HTL) and Upgrading," *Applied Energy* 129 (2014): 384-394, doi:https://doi.org/10.1016/j.apenergy.2014.03.053.

2.103. Demirbas, A., "Competitive Liquid Biofuels from Biomass," *Applied Energy* 88, no. 1 (2011): 17-28, doi:https://doi.org/10.1016/j.apenergy.2010.07.016.

2.104. Kourkoumpas, D.-S., Sagani, A., Hull, A., Hull, A. et al., "Life Cycle Assessment of an Innovative Alcohol-to-Jet Process: The Case for Retrofitting a Bioethanol Plant for Sustainable Aviation Fuel Production," *Renewable Energy* 228 (2024): 120512, doi:https://doi.org/10.1016/j.renene.2024.120512.

2.105. Richter, S., Kukkadapu, G., Westbrook, C.K., Braun-Unkhoff, M. et al., "A Combined Experimental and Modeling Study of Combustion Properties of an ISO-Paraffinic Alcohol-to-Jet Fuel," *Combustion and Flame* 240 (2022): 111994, doi:https://doi.org/10.1016/j.combustflame.2022.111994.

2.106. Prime Mover Labs, "Sustainable Aviation Fuel: Part 3, SAF Feedstocks and Production Costs," accessed November 4, 2024, https://medium.com/prime-movers-lab/sustainable-aviation-fuel-part-3-6d07f2a914ac.

2.107. Guzman, J., Kukkadapu, G., Brezinsky, K., and Westbrook, C., "Experimental and Modeling Study of the Pyrolysis and Oxidation of an ISO-Paraffinic Alcohol-to-Jet Fuel," *Combustion and Flame* 201 (2019): 57-64, doi:https://doi.org/10.1016/j.combustflame.2018.12.013.

2.108. Voß, S., Bube, S., and Kaltschmitt, M., "Aviation Fuel Production Pathways from Lignocellulosic Biomass via Alcohol Intermediates-A Technical Analysis," *Fuel Communications* 17 (2023): 100093, doi:https://doi.org/10.1016/j.jfueco.2023.100093.

2.109. PR Newswire, "Honeywell Revolutionizes Ethanol-to-Jet Fuel Technology to Meet Rising Demand for Sustainable Aviation Fuel," accessed December 21, 2024, https://www.prnewswire.com/news-releases/honeywell-revolutionizes-ethanol-to-jet-fuel-technology-to-meet-rising-demand-for-sustainable-aviation-fuel-301644079.html.

2.110. Gutiérrez-Antonio, C., Gómez-Castro, F.I., de Lira-Flores, J.A., and Hernández, S., "A Review on the Production Processes of Renewable Jet Fuel," *Renewable and Sustainable Energy Reviews* 79 (2017): 709-729, doi:https://doi.org/10.1016/j.rser.2017.05.108.

2.111. Hong, J., Chen, B., Wang, T., and Zhao, X., "A Promising Technical Route for Converting Lignocellulose to Bio-Jet Fuels Based on Bioconversion of Biomass and Coupling of Aqueous Ethanol: A Techno-Economic Assessment," *Fuel* 381 (2025): 133670, doi:https://doi.org/10.1016/j.fuel.2024.133670.

2.112. Su-ungkavatin, P., Tiruta-Barna, L., and Hamelin, L., "Biofuels, Electrofuels, Electric or Hydrogen?: A Review of Current and Emerging Sustainable Aviation Systems," *Progress in Energy and Combustion Science* 96 (2023): 101073, doi:https://doi.org/10.1016/j.pecs.2023.101073.

2.113. Chiaramonti, D., "Sustainable Aviation Fuels: The Challenge of Decarbonization," *Energy Procedia* 158 (2019): 1202-1207, doi:https://doi.org/10.1016/j.egypro.2019.01.308.

2.114. Wang, M., Dewil, R., Maniatis, K., Wheeldon, J. et al., "Biomass-Derived Aviation Fuels: Challenges and Perspective," *Progress in Energy and Combustion Science* 74 (2019): 31-49, doi:https://doi.org/10.1016/j.pecs.2019.04.004.

2.115. Wei, H., Liu, W., Chen, X., Yang, Q. et al., "Renewable Bio-Jet Fuel Production for Aviation: A Review," *Fuel* 254 (2019): 115599, doi:https://doi.org/10.1016/j.fuel.2019.06.007.

2.116. Marchesan, A.N., de Mesquita Sampaio, I.L., Chagas, M.F., Generoso, W.C. et al., "Alternative Feedstocks for Sustainable Aviation Fuels: Assessment of Sugarcane-Derived Microbial Oil," *Bioresource Technology* 416 (2025): 131772, doi:https://doi.org/10.1016/j.biortech.2024.131772.

2.117. Prussi, M., Weindorf, W., Buffi, M., Sánchez López, J. et al., "Are Algae Ready to Take Off? GHG Emission Savings of Algae-to-Kerosene Production," *Applied Energy* 304 (2021): 117817, doi:https://doi.org/10.1016/j.apenergy.2021.117817.

2.118. Monteiro, R.R.C., dos Santos, I.A., Arcanjo, M.R.A., Cavalcante, C.L. Jr. et al., "Production of Jet Biofuels by Catalytic Hydroprocessing of Esters and Fatty Acids: A Review," *Catalysts* 12, no. 2 (2022): 237, doi:https://doi.org/10.3390/catal12020237.

2.119. Vásquez, M.C., Silva, E.E., and Castillo, E.F., "Hydrotreatment of Vegetable Oils: A Review of the Technologies and Its Developments for Jet Biofuel Production," *Biomass and Bioenergy* 105 (2017): 197-206, doi:https://doi.org/10.1016/j.biombioe.2017.07.008.

2.120. Yáñez, É., Meerman, H., Ramírez, A., Castillo, É. et al., "Assessing Bio-Oil Co-Processing Routes as CO_2 Mitigation Strategies in Oil Refineries," *Biofuels, Bioprod. Bioref.* 15 (2021): 305-333, doi:https://doi.org/10.1002/bbb.2163.

2.121. van Dyk, S., Su, J., Mcmillan, J.D., and Saddler, J., "Potential Synergies of Drop-In Biofuel Production with Further Co-Processing at Oil Refineries," *Biofuels, Bioprod. Bioref.* 13 (2019): 760-775, doi:https://doi.org/10.1002/bbb.1974.

2.122. Bezergianni, S., Dimitriadis, A., Kikhtyanin, O., and Kubička, D., "Refinery Co-Processing of Renewable Feeds," *Progress in Energy and Combustion Science* 68 (2018): 29-64, doi:https://doi.org/10.1016/j.pecs.2018.04.002.

2.123. Li, D., Meng, X., Huang, S., Wang, J. et al., "Aromatization-Free Strategy for Sustainable Aviation Fuel: Co-Upcycling Waste Oils and Polystyrene Plastics into Alkanes and Aromatics," *Chemical Engineering Science* 304 (2025): 121100, doi:https://doi.org/10.1016/j.ces.2024.121100.

2.124. Biomass Research Development, "SAF Grand Challenge Roadmap: Conversion Technology Innovation Action Area," accessed December 1, 2024, https://biomassboard.gov/saf-grand-challenge-roadmap-conversion-technology-innovation-action-area.

2.125. The National Aeronautics and Space Administration (NASA), "Technology Readiness Levels," accessed December 6, 2024, https://www.nasa.gov/directorates/somd/space-communications-navigation-program/technology-readiness-levels/.

2.126. Goh, B.H.H., Chong, C.T., Ong, H.C., Seljak, T. et al., "Recent Advancements in Catalytic Conversion Pathways for Synthetic Jet Fuel Produced from Bioresources," *Energy Conversion and Management* 252 (2022): 114974, doi:https://doi.org/10.1016/j.enconman.2021.114974.

2.127. O'Connell, A., Kousoulidou, M., Lonza, L., and Weindorf, W., "Considerations on GHG Emissions and Energy Balances of Promising Aviation Biofuel Pathways," *Renewable and Sustainable Energy Reviews* 101 (2019): 504-515, doi:https://doi.org/10.1016/j.rser.2018.11.033.

2.128. Roland Berger, "Power-to-Liquid Fuels Will Be Crucial to Power Large Long-Haul Aircraft," accessed December 3, 2024, https://www.rolandberger.com/en/Insights/Publications/Sustainable-aviation-fuels-key-for-the-future-of-air-travel.html.

2.129. Yilmaz, N. and Atmanli, A., "Sustainable Alternative Fuels in Aviation," *Energy* 140, no. 2 (2017): 1378-1386, doi:https://doi.org/10.1016/j.energy.2017.07.077.

2.130. Zhang, H., Fang, Y., Wang, M., Appels, L. et al., "Prospects and Perspectives Foster Enhanced Research on Bio-Aviation Fuels," *Journal of Environmental Management* 274 (2020): 111214, doi:https://doi.org/10.1016/j.jenvman.2020.111214.

2.131. Choi, I.-H., Hwang, K.-R., Han, J.-S., Lee, K.-H. et al., "The Direct Production of Jet-Fuel from Non-Edible Oil in a Single-Step Process," *Fuel* 158 (2015): 98-104, doi:https://doi.org/10.1016/j.fuel.2015.05.020.

2.132. Detsios, N., Theodoraki, S., Maragoudaki, L., Atsonios, K. et al., "Recent Advances on Alternative Aviation Fuels/Pathways: A Critical Review," *Energies* 16, no. 4 (2023): 1904, doi:https://doi.org/10.3390/en16041904.

2.133. Bekele, E., Sgaramella, A., Ciancio, A., Basso, G.L. et al., "Power-to-Gas and Power-to-Liquid Systems in Emerging Hydrogen Valleys: Techno-Economic Assessment of Alternative Fuels," *Energy Storage and Saving* 4, no. 2 (2025): 148-156, doi:https://doi.org/10.1016/j.enss.2025.01.003.

2.134. Rojas-Michaga, M.F., Michailos, S., Cardozo, E., Hughes, K.J. et al., "A Techno-Economic and Life Cycle Assessment of a New Power and Biomass to Liquids (PBtL) Configuration with Negative Emissions for Producing Sustainable Aviation Fuel (SAF)," *Energy Conversion and Management: X* 25 (2025): 100841, doi:https://doi.org/10.1016/j.ecmx.2024.100841.

2.135. LBST – Ludwig-Bölkow-Systemtechnik GmbH, "E-SAF: Techno-Economics of PtL and PtH2," accessed December 8, 2024, https://lbst.de/wp-content/uploads/2023/12/DA_E-SAF_Report_final_2023_12_04.pdf.

2.136. ETHA Zürich, "Fuels from Sunlight and Air," accessed January 4, 2025, https://prec.ethz.ch/ffsaa.html.

2.137. Romero, M. and Steinfeld, A., "Concentrating Solar Thermal Power and Thermochemical Fuels," *Energy & Environmental Science* 5 (2012): 9234-9245, doi:https://doi.org/10.1039/C2EE21275G.

2.138. Dähler, F., Wild, M., Schäppi, R., Haueter, P. et al., "Optical Design and Experimental Characterization of a Solar Concentrating Dish System for Fuel Production via Thermochemical Redox Cycles," *Solar Energy* 170 (2018): 568-575, doi:https://doi.org/10.1016/j.solener.2018.05.085.

2.139. De Cata, D., Mazzeo, L., Piemonte, V., and Giaconia, A., "Electrified Steam Methane Reforming as Efficient Pathway for Sustainable Hydrogen Production and Industrial Decarbonization: A Critical Review," *International Journal of Hydrogen Energy* 105 (2025): 31-44, doi:https://doi.org/10.1016/j.ijhydene.2025.01.202.

2.140. Kazula, S., "Common Cause Analysis of the Air Supply System of Fuel Cell-Powered Propulsion Systems in Electrified Aviation," *Engineering Failure Analysis* 163, no. A (2024): 108435, doi:https://doi.org/10.1016/j.engfailanal.2024.108435.

2.141. Sustainable Skies, "Hydrogen Malibu's on Two Continents," accessed December, 28 2024, https://sustainableskies.org/hydrogen-malibus-on-two-continents/.

2.142. Zeroavia, "Hydrogen Fuel for Zero-Emission Flight," accessed December 12, 2024, https://zeroavia.com/newsroom/.

2.143. Massaro, M.C., Biga, R., Kolisnichenko, A., Marocco, P. et al., "Potential and Technical Challenges of On-Board Hydrogen Storage Technologies Coupled with Fuel Cell Systems for Aircraft Electrification," *Journal of Power Sources* 555 (2023): 232397, doi:https://doi.org/10.1016/j.jpowsour.2022.232397.

2.144. Energy Industry Review, "World's First Hydrogen-Based, 1.5 MW Propulsion System for Aeronautical Applications," accessed January 2, 2025, https://energyindustryreview.com/energy-efficiency/worlds-first-hydrogen-based-1-5-mw-propulsion-system-for-aeronautical-applications/.

2.145. Zhang, L., Jia, C., Bai, F., Wang, W. et al., "A Comprehensive Review of the Promising Clean Energy Carrier: Hydrogen Production, Transportation, Storage, and Utilization (HPTSU) Technologies," *Fuel* 355 (2024): 129455, doi:https://doi.org/10.1016/j.fuel.2023.129455.

2.146. López-Aranguren, P., Lannelongue, P., Carrasco, J., Golov, A. et al., "HELENA Project: Driving Innovation in High Energy Density Li-Metal Halide Solid-State Batteries for Electric Vehicles and Aircrafts," *Computational and Structural Biotechnology Journal* 29 (2025): 72-84, doi:https://doi.org/10.1016/j.csbj.2025.03.013.

2.147. Vališ, D., Hlinka, J., Koláček, J., Procházka, P. et al., "Comprehensive Study on Reliability, Deterioration, and Ageing of Lithium-Ion Batteries in Electric-Powered Aircraft Assessed after a Storage Period Using Functional Data Analysis and Non-Parametric Back Up," *Engineering Failure Analysis* 154 (2023): 107661, doi:https://doi.org/10.1016/j.engfailanal.2023.107661.

Chapter 03

Environmental Impact of SAFs

As global trade and economic growth expand, the demand for aviation fuel in the airline transportation industry will continue to increase in the coming years. The climate impact of aviation is attributable to direct aircraft emissions and the resulting atmospheric effects that occur as a consequence of those emissions [3.1]. The primary goal of promoting SAF adoption and implementing legal mandates is to reduce the aviation industry's carbon footprint and support its transition toward net-zero emissions, fostering a more sustainable and environmentally friendly future. Renewable fuel mandates in the aviation industry have always been difficult due to debates over how to distribute GHG reductions based on flight departures and destinations. The ICAO currently has an SAF program that will officially come into force in 2027, but some countries are progressing faster to set national targets [3.2]. The supply of biomass, waste, and other raw materials required for liquid fuels depends on land, water, and clean energy, which is assessed within the scope of SAFs. Environmental safety is a fundamental element of the aviation industry, and it is evaluated collectively in terms of soil, air, water, and human health, as shown in **Figure 3.1** [3.3].

Figure 3.1 Environmental pollution [3.3].

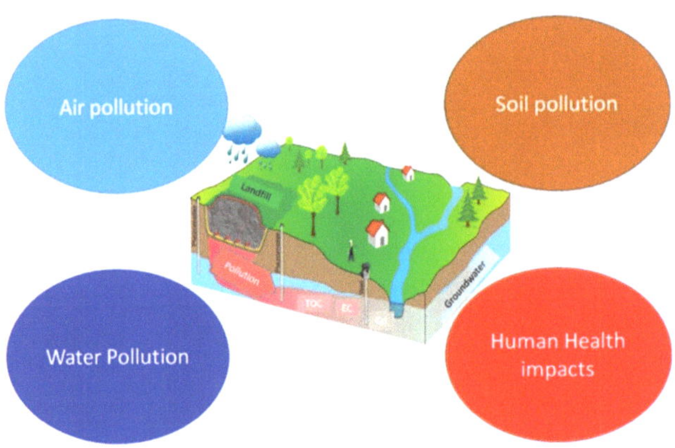

Processing waste oils and animal fats is subject to pollution from higher levels of contaminants such as nitrogen, metals, and polyethylene, compared to virgin oil feeds [3.4]. These contaminants must be removed to maximize product yield and catalyst cycle length. As a result, renewable fuel processing technologies have emerged that include pretreatment methods to effectively neutralize contaminants that may originate from raw feedstocks. In order to effectively manage the production, supply, and fuel production processes of raw feedstocks, these environmental factors must be controllable at every stage [3.5]. The Roundtable on Sustainable Biomaterials (RSB) is a global organization dedicated to advancing the transition to a biobased and circular economy. Sustainability, in this context, refers to preserving the long-term functionality of systems. The RSB's Principles and Criteria (P&C) framework outlines key sustainability benchmarks, addressing critical issues in SAF production and the broader bioeconomy [3.6]. The key factors in the implementation of successful public policies to shape the sustainable future of the aviation industry are outlined in **Figure 3.2** [3.7].

Figure 3.2 A white paper of the RSB SAF policy [3.7].

Evaluating these 12 principles in an environmental context during the SAF production process can play an important role in combating the threat of climate change. Aviation industry participants have committed to halving net CO_2 emissions by 2050 [3.8]. Transitioning from fossil fuels to SAFs is considered one of the most effective ways to achieve this decarbonization target within the stipulated time. If the SAF industry meets these criteria, it can move beyond decarbonization to foster the development of a biobased and circular economy. It has the potential to protect and enhance rural livelihoods, communities, and ecosystems, as well as increasing the use of new raw materials and new supply chains that can develop new technologies [3.9]. When evaluated in terms of future projection, SAFs are quite low worldwide with the current production infrastructure. They currently account for less than 0.01% of all jet fuel produced worldwide. According to current Energy Information Administration (EIA) data, aviation fuel is identified as the fastest-growing fuel segment, with global demand expected to grow from 106 billion gallons in 2019 to 230 billion gallons in 2050 [3.8, 3.9]. Renewable jet fuel production is slightly more expensive than renewable diesel production. This is due to additional hydrocracking and increased levels of hydrogen consumption by jets. Capital costs are also slightly higher due to additional processing equipment such as the need for a vapor stripper, separation column, and separated product storage, as well as additional laboratory and certification requirements. SAF production needs to be increased rapidly to achieve the volumes and prices that are required to be competitive. Similar to other renewable fuels such as ethanol and biodiesel, this rapid scaling is achievable, but a supportive policy environment is vital [3.10]. There are many practical methods to reduce the emission of GHGs in aviation, such as structural equipment changes in advanced aircraft, but supporting the use of SAFs currently offers an environmentally effective solution. Sustainable aviation and climate resilience are fundamental to today's aviation industry. In this context, many countries that monitor the use of SAFs with policy are currently adopting or proposing additional guidelines to increase the use of SAFs. However, it is vital that these policies are supported by social and environmental sustainability principles. The initiatives taken by airports and airlines to reduce their impact on climate change and on themselves will define the environmental sustainability of the industry for decades to come. SAF production is determined by two main factors: the raw material used and the technology used to convert the raw material into HCs [3.11]. Both factors have an impact on the economic and environmental sustainability of production and, as a result, on the supply, widespread use, and price of SAFs on the market. As shown in **Figure 3.3**, it is important to plan useful energy consumption and reduce carbon emissions as a result of energy production and use [3.12].

Figure 3.3 Energy production and useful energy for SAFs [3.12].

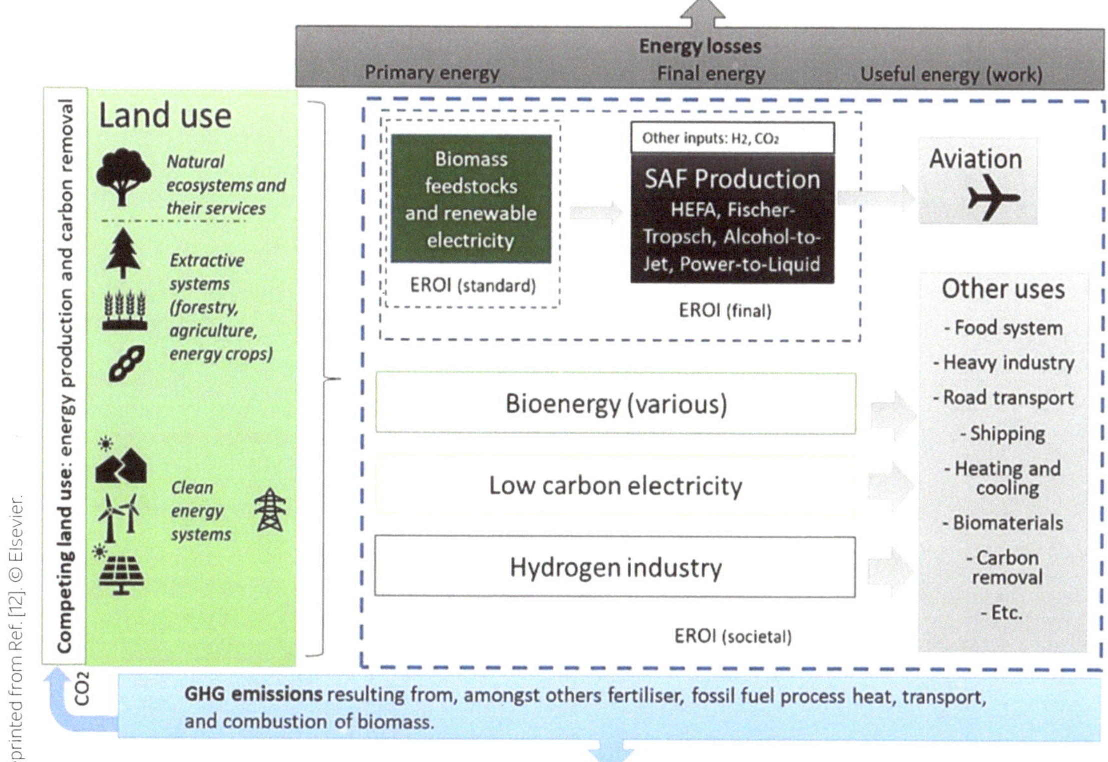

Roadmaps are being developed to achieve these useful energy targets, which are determined by evaluating all aspects of SAF use in the aviation industry. Aviation roadmaps are created following extensive consultations with industry, academia, aviation, and international stakeholders. Significant policy developments, such as the "ReFuelEU" regulation to introduce the production and use of SAFs, also address the environmental impacts of fuels [3.13]. The European Commission and the European Union Aviation Safety Agency (EASA) are also working in collaboration with the aviation industry to develop an environmental label for the aviation sector so that passengers can access reliable environmental information. These organizations can determine the decision-making process between different options in the sector by evaluating the performance of aircraft, airlines, and individual flights and complementing current roadmaps for net-zero emission targets [3.14]. Aviation's impact on the climate is complex, extending beyond fossil CO_2 emissions to include several non-CO_2 effects that contribute to warming when combined. In some cases, research and development on non-CO_2 mitigation technologies is integrated into existing technology roadmaps. An analysis of 12 aviation roadmaps for achieving net-zero emissions by 2050 shows a strong reliance on SAFs in the medium term, with synthetic e-kerosene expected to play a larger role in the long term. Realizing these roadmaps could require 9% of global renewable electricity and 30% of sustainable biomass by 2050, representing significant progress [3.15]. According to roadmap projections, if the use of HC fuels continues, 1.35 $GtCO_2$ will be produced in 2050, and 30% of this will still be provided by fossil fuels [3.13, 3.15]. In the case of using SAFs, 70% net carbon saving depends on the direct and indirect lifecycle emissions based on the processes in SAF production. With the widespread use of biomass-based SAFs, a large projected reduction in CO_2 emissions is predicted, as shown in **Figure 3.4** [3.16].

The use of large amounts of land for biomass production and the consumption of water and other natural resources are taken into consideration while creating roadmaps to ensure environmental sustainability in energy production. Similarly, processing waste may present challenges due to high levels of contaminants such as nitrogen, metals, and polyethylene compared to raw materials containing virgin oil [3.17]. These contaminants must be removed to maximize product yield and catalyst cycle length. This indicates the pretreatment of feedstock and a renewable fuel process technology that can work with contaminants.

Figure 3.4 Projection of net-zero carbon emissions by the aviation sector by 2050 [3.16].

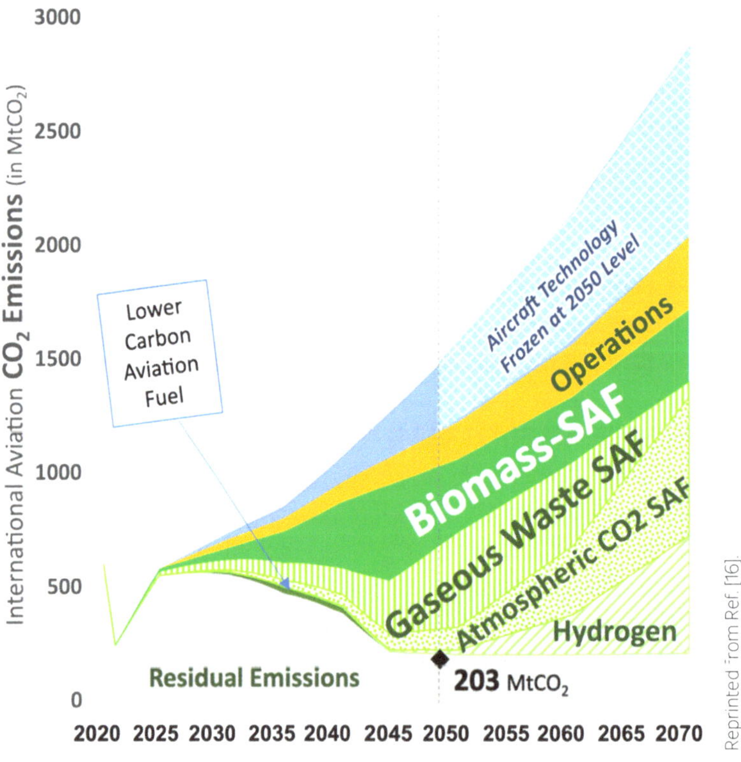

SAF Roadmaps for Pollution Reduction

Roadmaps are the first detailed assessment of the key steps needed to accelerate the transition to net zero by 2050. Together, they provide a clear direction and will evolve as we dig deeper to identify intermediate milestones on the path to net zero. Roadmaps are aimed at airlines, but also governments, suppliers, and financiers [3.18]. The environmental impacts of any roadmap and policy for SAFs should be detailed. The main aim of the roadmaps planned in this context is to reduce CO_2 emissions to the targeted level. In addition to CO_2 emission reduction targets, a total emission reduction roadmap based on emissions (residual emissions) other than CO_2 is becoming more important in terms of environmental sustainability [3.19]. In general, non-CO_2 emissions refer to the direct and indirect effects of combustion, as well as CO_2 found in aircraft exhaust fumes. Exhaust gases often contain CO_2 and other pollutants, with specifics dependent on the engine and fuel type. The roadmap shown in **Figure 3.5** offers the most effective example of this in terms of its scope [3.20].

Figure 3.5 Non-CO_2 technologies roadmap [3.20].

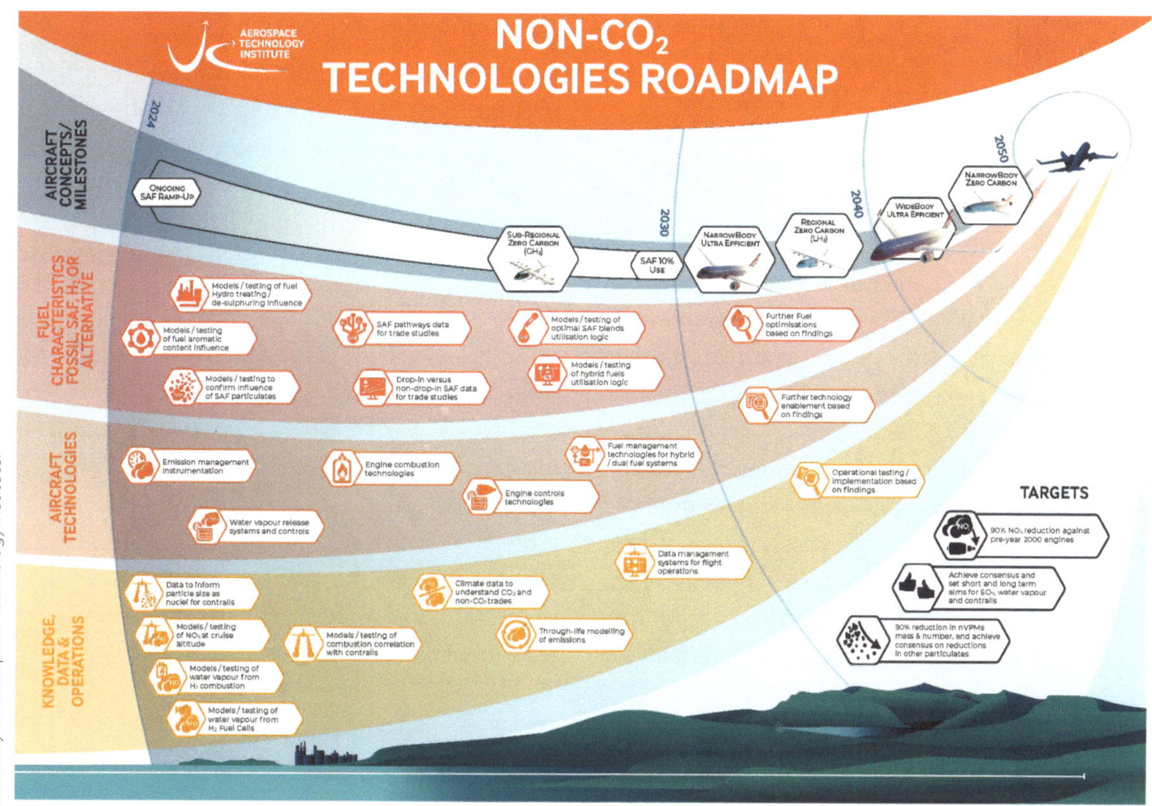

Courtesy of Aerospace Technology Institute.

The non-CO_2 technologies roadmap, one of the roadmaps that offers an environmental and technological approach, is in force as a pillar of the UK Aerospace Technology Strategy for net zero. Addressing the priority areas identified in the non-CO_2 technologies roadmap can reduce environmental uncertainty and provide technology solutions to reduce or prevent non-CO_2 emissions in flight. In addition to roadmaps on a country basis, aircraft-manufacturing companies are also developing their own approaches [3.21].

Airbus's perspective of the net-zero target includes energy-related technologies such as incremental improvements with SAFs, new hybrid electric engine design, and future electric flight [3.22]. All Airbus aircraft are certified to operate on up to a 50% SAF blend [3.23]. Due to current engine technology and fuel infrastructure, liquid fuels will continue to be used in jet engines. The 100% SAF project is ongoing in line with this roadmap, which will help support future SAF certification for mixtures exceeding the current maximum of 50%. In addition to biomass-based fuel, other alternative fuels such as PtL synthetic e-fuel made from renewable hydrogen and CO_2, obtained by capturing carbon directly from air, represent a major opportunity to reduce aviation CO_2 emissions [3.24]. While efficiency and operational improvements can reduce GHG emissions, decarbonization must rely heavily on low-carbon kerosene alternatives, namely SAFs. SAFs compete for land needed for nature-based decarbonization, clean energy that can decarbonize other sectors more effectively, and CO_2 captured for permanent storage. The biomass supply phase is an important factor for SAFs to not only sustain but also expand and scale up global aviation [3.25]. As the demand for fuel products changes, countries are looking for ways to reuse existing assets to produce renewable fuels efficiently. The emergence of new technologies to use sustainable feeds and improve them, such as low-quality residual biomass, MSW streams, and recycled carbon, can reduce carbon emissions. This also allows the industry to introduce new aircraft and fuel types with minimal disruption to existing operations [3.24, 3.25]. Intensive work is being done on further solutions toward net-zero-emission air travel. Whether innovative aircraft designs or novel fuels, all solutions have two things in common: they are cost-intensive, and it will take many years before the developments are ready for the market and emission-free flying becomes a reality [3.26]. As an environmental protection strategy of the international aviation industry, investments have been made over many years in order to reduce the amount of fuel consumed per passenger and therefore reduce emissions [3.27].

Despite COVID-19 and the current restrictions imposed by the global security situation, instead of decreasing air traffic, air passenger volume is expected to increase by 3.7% per year [3.18, 3.19]. In this increasing traffic, only CO_2 emissions are taken into account in the 2050 target. However, due to the operating principle of aircraft engines, post-combustion emission treatment systems are not as strict as in land vehicles. Non-CO_2 emissions from aviation can also contribute to the climate impact in terms of global warming [3.25].

Kerosene-based aviation fuels lead to emissions of other harmful pollutants (**Figure 3.6**) as follows:

- PM, soot, and unburnt HCs: SAFs have been shown to reduce PM emissions significantly due to their lower sulfur and aromatic content [3.29].
- SO_X: Sulfur compounds, including sulfur oxides (SO_X) and sulfate aerosols in SAFs, typically contain very low sulfur levels, leading to a substantial reduction in SO_X emissions, which are associated with acid rain and respiratory problems [3.24, 3.29].
- CO: Thanks to the oxygen content of SAFs, CO formation is prevented.
- NO_x: Although NO_x emissions are more dependent on combustion temperature and engine conditions, some studies suggest that SAFs can contribute to modest reductions in NO_x emissions due to their cleaner combustion properties [3.26, 3.29].

The climate impact of aircraft emissions is extremely complex and varies depending on many factors, including the altitude dependence of emissions, weather patterns, and regional characteristics such as the presence of ice-supersaturated regions. The effects of residue emissions on climate are depicted in **Figure 3.7** [3.30]. When the environmental effects of the pollutants mentioned due to the use of conventional fuels are examined, the amount of fuel consumed and engine performance increase the emissions of these pollutants. As the aviation industry aims for net-zero CO_2 emissions by 2050, the environmental impacts of other emissions, including nonvolatile particulate matter (nvPM) and NO_x, also need to be addressed.

Figure 3.6 Aircraft emissions and environmental impacts [3.28].

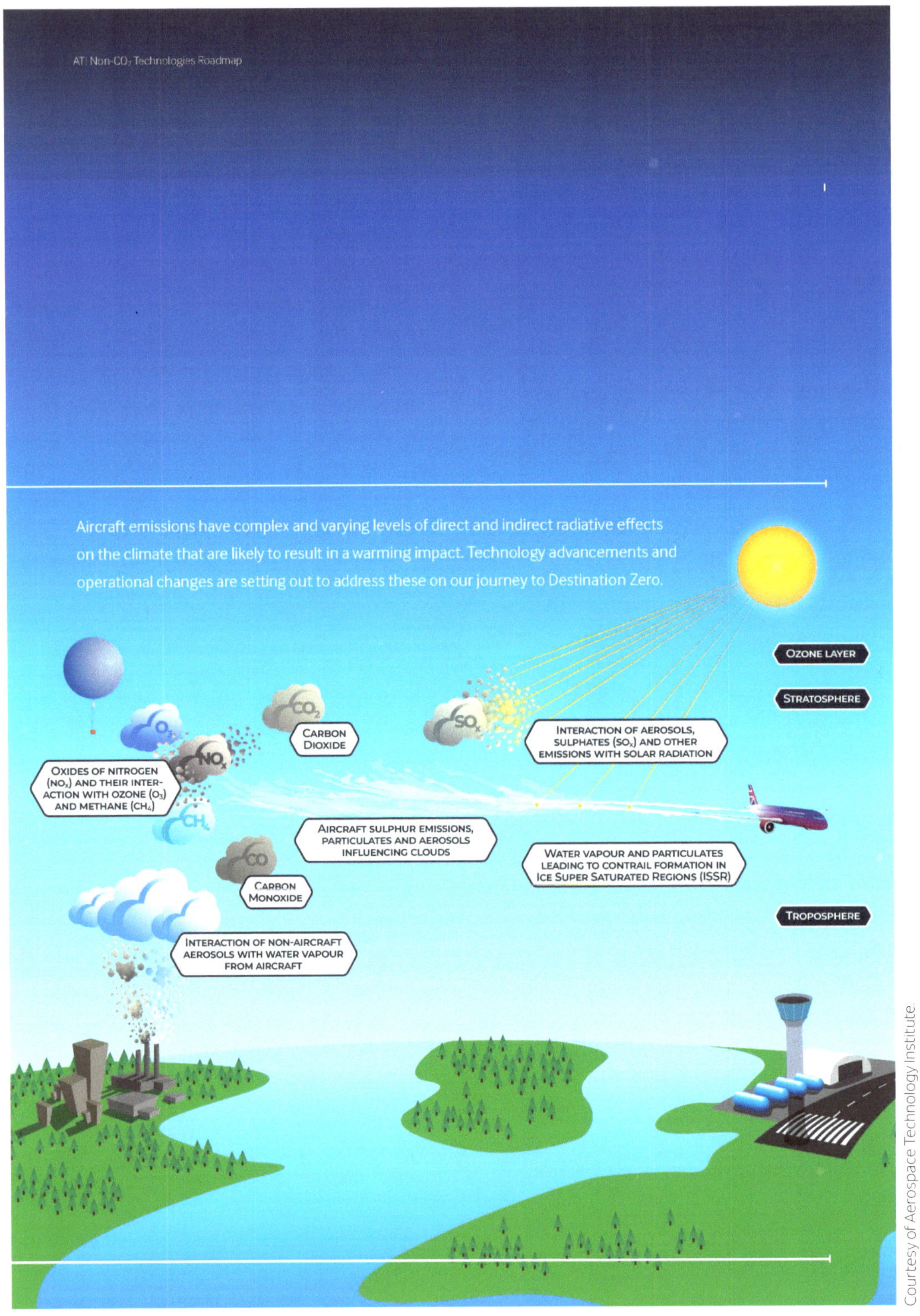

Courtesy of Aerospace Technology Institute.

Figure 3.7 Comparison of CO_2 and non-CO_2 emissions [3.30].

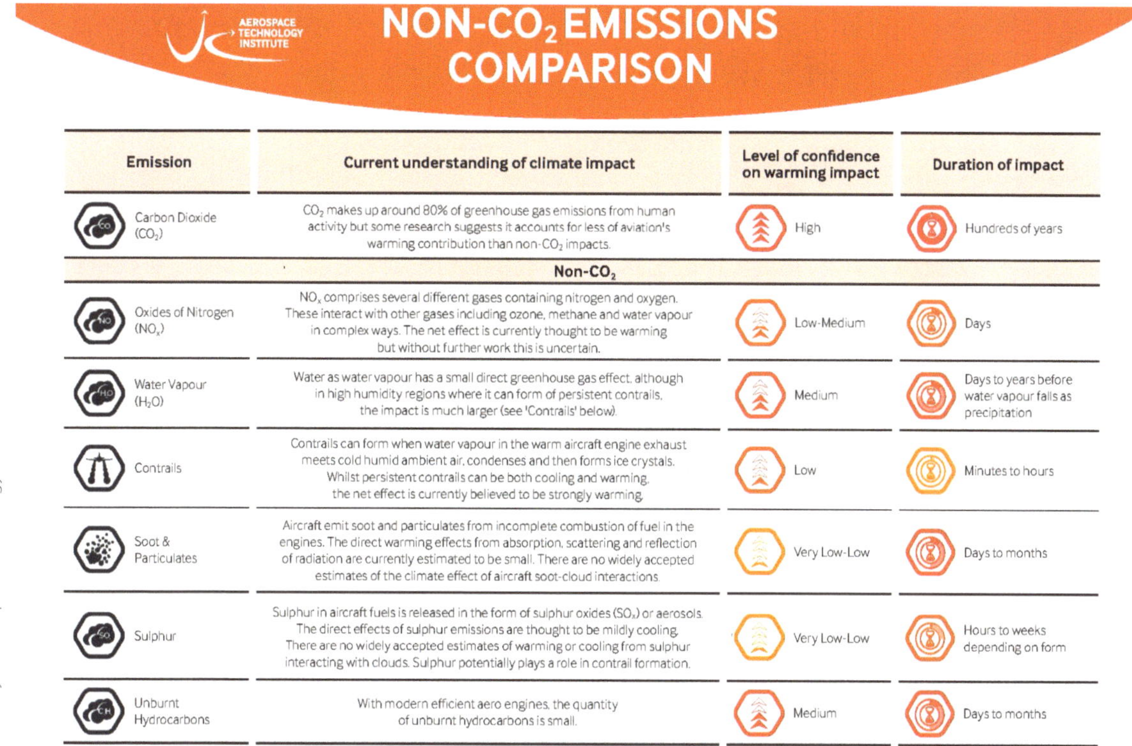

Emission	Current understanding of climate impact	Level of confidence on warming impact	Duration of impact
Carbon Dioxide (CO_2)	CO_2 makes up around 80% of greenhouse gas emissions from human activity but some research suggests it accounts for less of aviation's warming contribution than non-CO_2 impacts.	High	Hundreds of years
Non-CO_2			
Oxides of Nitrogen (NO_x)	NO_x comprises several different gases containing nitrogen and oxygen. These interact with other gases including ozone, methane and water vapour in complex ways. The net effect is currently thought to be warming but without further work this is uncertain.	Low-Medium	Days
Water Vapour (H_2O)	Water as water vapour has a small direct greenhouse gas effect, although in high humidity regions where it can form of persistent contrails, the impact is much larger (see 'Contrails' below).	Medium	Days to years before water vapour falls as precipitation
Contrails	Contrails can form when water vapour in the warm aircraft engine exhaust meets cold humid ambient air, condenses and then forms ice crystals. Whilst persistent contrails can be both cooling and warming, the net effect is currently believed to be strongly warming.	Low	Minutes to hours
Soot & Particulates	Aircraft emit soot and particulates from incomplete combustion of fuel in the engines. The direct warming effects from absorption, scattering and reflection of radiation are currently estimated to be small. There are no widely accepted estimates of the climate effect of aircraft soot-cloud interactions.	Very Low-Low	Days to months
Sulphur	Sulphur in aircraft fuels is released in the form of sulphur oxides (SO_x) or aerosols. The direct effects of sulphur emissions are thought to be mildly cooling. There are no widely accepted estimates of warming or cooling from sulphur interacting with clouds. Sulphur potentially plays a role in contrail formation.	Very Low-Low	Hours to weeks depending on form
Unburnt Hydrocarbons	With modern efficient aero engines, the quantity of unburnt hydrocarbons is small.	Medium	Days to months

Courtesy of Aerospace Technology Institute.

Experts from the industry, government, and academia reveal a significant projection for reducing the environmental impacts of aviation, including impacts caused by non-CO_2 emissions, as shown in **Figure 3.8** [3.31]. CO_2 and non-CO_2 emissions should first be kept under control within the targeted limits and then reduced using sustainable roadmaps. Using basic policy development and solution stages, improving fuel and aircraft engine/power unit technology and choosing a common path by evaluating climatic parameters together will advance the roadmaps to success. Research on SAFs and other alternative fuels, including hydrogen in both FCs and gas turbine combustion, and increasing the current use of SAFs needs to be supported. At the same time, adaptations to the kerosene specification to reduce the harmful effects of non-CO_2 emissions, as well as optimized operational use of hybrid fuel combinations or SAFs, stand out in fuel-related methods.

Figure 3.8 All emissions and different roadmap solutions in the aviation industry [3.31].

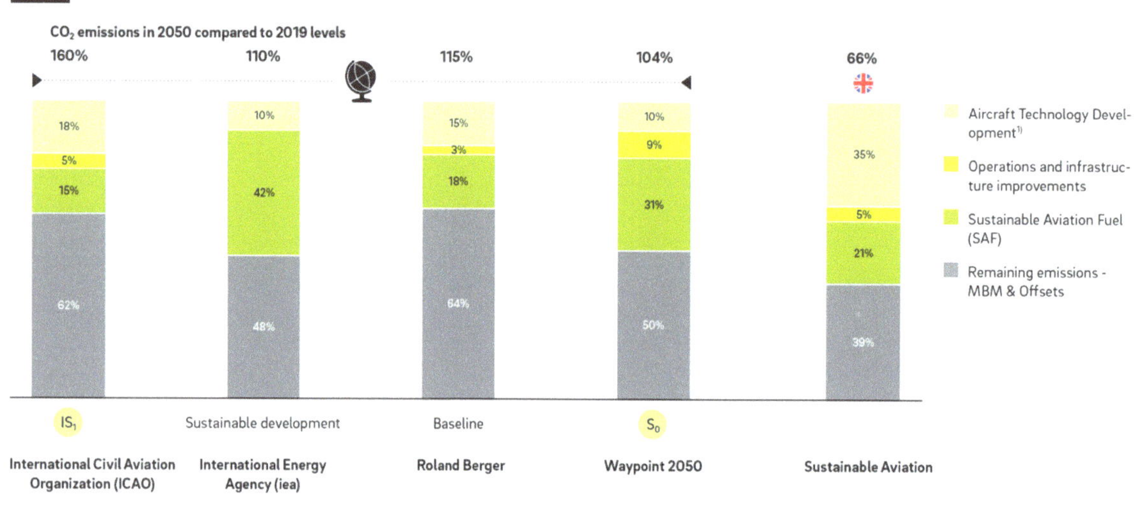

Instrumentation and emission detection solutions should be developed for research, emission measurement, and management at the feeding stage of aircraft engines and power units [3.32]. Systems and controls for the use of hydrogen, including water vapor (H_2O) release, in terms of engine technology; engine controls and combustion improvements; and technologies for managing hybrid and dual-fuel systems need to be developed. Thus, a multifaceted approach needs to be implemented for all pollutants that will arise from the use of SAFs. To truly realize the aim of protecting the environment, it is important to raise awareness of keeping both CO_2 and non-CO_2 emissions under control in all future projections and roadmaps, including those that support using SAFs [3.33].

SAFs for Emission Reduction Policy

The contribution of SAF use to reducing emissions has been acknowledged by countries, airlines, and scientists working on the issue worldwide. The positive effects of alternative fuels on non-CO_2 emissions have been recognized. Current research on climate science, meteorology, and engine emissions provides a good understanding of the non-CO_2 effects of alternative fuels such as SAFs for integrated climate impact assessments [3.34]. With this positive thinking and developmental infrastructure, aviation is experiencing transformative change. Many converging factors are accelerating the drive toward sustainable aviation. One of the main drivers is the increasing pressure to reduce the environmental impact of the aviation industry. Governments, regulators, and local communities demand more sustainable practices from airports. This legitimate demand of end-user passengers has been positively reflected in the aviation ecosystem, both in advertisements and in newly constructed airports (**Figure 3.9**) [3.35]. Sustainable environmental protection can be achieved if airports are powered by clean energy and aircraft operate on sustainable fuels like SAFs.

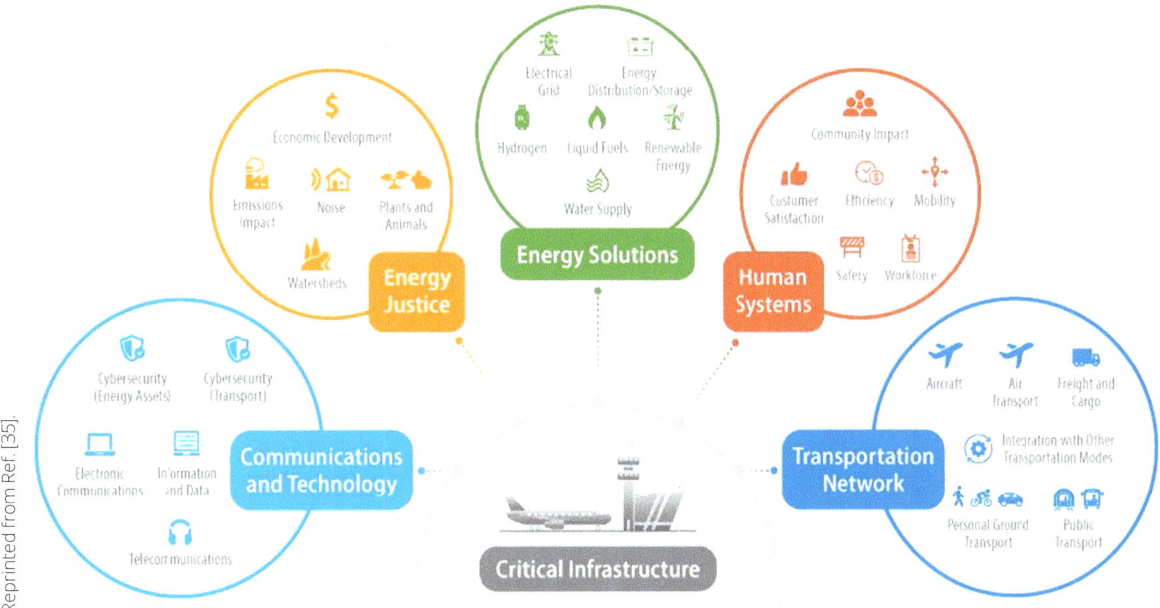

Figure 3.9 Airport ecosystem [3.35].

Reprinted from Ref. [35].

As the aviation industry expands, the environmental and climatic impacts of aviation emissions become critically important. Emissions that need to be kept under control are supported by emission reduction roadmaps and emission regulations such as net zero, the effectiveness of the Renewable Energy Directive III (RED III) and CORSIA schedules, and the use of SAFs, with additional policies from ReFuelEU and Fitfor55 [3.36].

RED II covers the period up to 2030, and its overarching goal is to achieve at least 32% of Europe's energy from renewable sources by 2030. RED II contains sustainability and GHG-saving criteria for biomass to be used as feedstock for the production of transport fuels or to be converted into electricity, heating, or cooling. However, it adopted an amendment in 2023, called "RED III," which significantly raised the collective target for renewable energy consumption across all sectors in Europe to at least 42.5% by 2030 [3.37]. At the same time, new accompanying measures are required in different sectors in line with the integration of the energy system, hydrogen, offshore renewable energy, and biodiversity strategies to achieve this modified goal.

In 2010, the ICAO decided to achieve carbon-neutral growth. In 2016, it adopted the CORSIA, requiring airlines to purchase carbon offset credits for emissions above 2019 levels on international flights from 2021 onward [3.38]. Carbon offset projects are defined as those that reduce or store CO_2 or other GHG emissions, for example, by investing in renewable energy or forest conservation to offset emissions elsewhere. In the aviation sector, carbon offsetting policy is seen as a potential second-best solution because sectoral marginal abatement costs are relatively high and potential best solutions such as carbon taxes and cap-and-trade systems have not proven politically feasible. Since ICAO is responsible for civil operations only, military and government flights are completely excluded. Monitoring, reporting, and verification (MRV) obligations began in 2019 [3.39]. As shown in **Figure 3.10**, under CORSIA, MRV of CO_2 emissions is mandatory for all ICAO contracting states, regardless of voluntary or compulsory participation [3.40]. CORSIA consists of three phases: Pilot Phase (2021–2023), Phase 1 (2024–2026), and Phase 2 (2027–2035). While Pilot Phase and Phase 1 are voluntary, participation in Phase 2 is mandatory for all states with airlines that account for more than 0.5 of the global international revenue ton kilometers (RTK) in 2018. Only least developed countries, landlocked developing countries, and small island states are exempt from this obligation. Of course, voluntary participation is possible for any ICAO contracting state. Balancing requirements are defined according to the routes flown [3.39, 3.40]. The continued compliance with CORSIA is a sign of increased climate and regulatory integrity. However, some challenges remain in achieving carbon-neutral growth. In particular, the use of SAFs needs to be significantly promoted to reduce fuel emissions in air transport.

The success of CORSIA will depend largely on the offsetting measures used. To ensure maximum climate and regulatory integrity of CORSIA in the aviation sector, policymakers should consider prioritizing relatively small carbon reduction projects and selected carbon storage projects targeting energy efficiency at the consumer level. In order to achieve all the determined targets, a triple integration approach needs to be achieved in the sector [3.41].

Thus, CORSIA is supported by several major carbon reduction schemes, such as the US, EU, and UK Emissions Trading Scheme (ETS), and could become more popular as a way to offset carbon reductions by improving energy efficiency [3.42]. CORSIA's eligibility criteria need to be regularly improved to get closer to the carbon-neutral growth aspiration. Airlines should also be encouraged to provide clean flights and invest more in sustainable and advanced technologies to reduce environmental impacts and meet emission targets across sectors [3.43].

Figure 3.10 Design of the CORSIA [3.40].

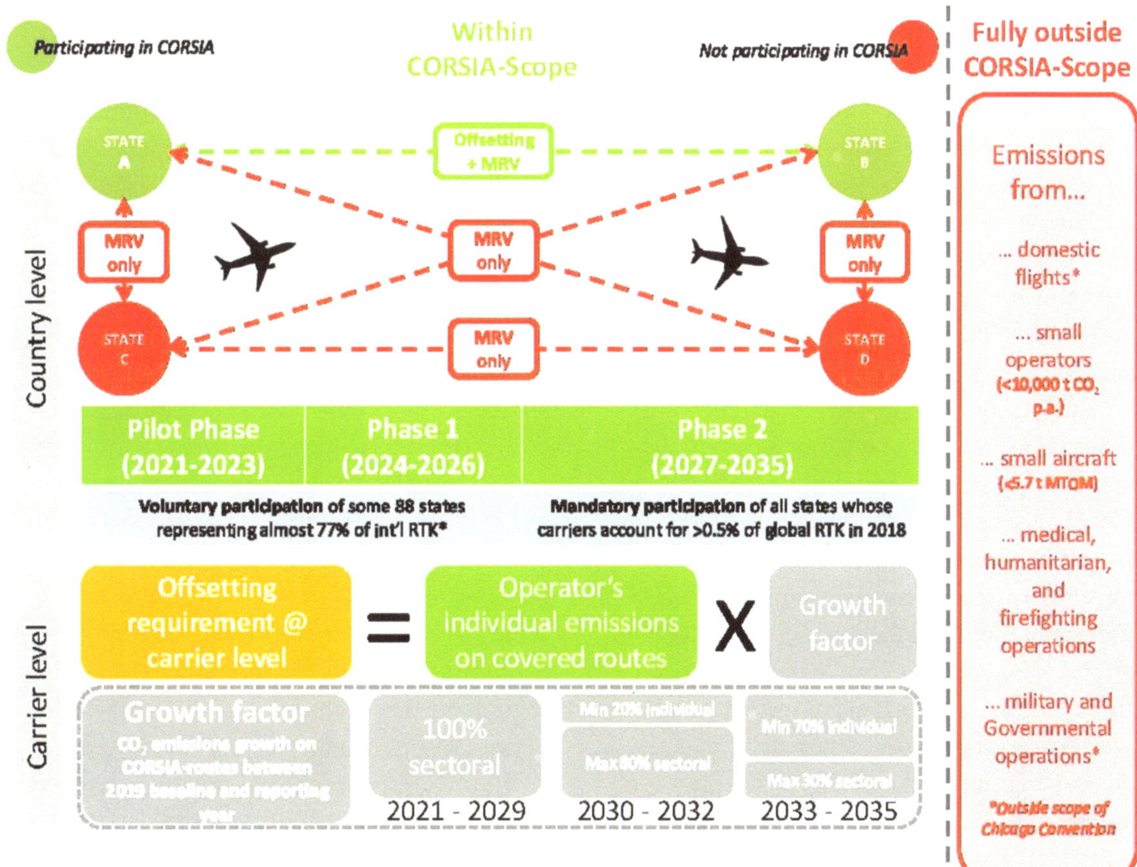

Combustion Parameters of SAFs

In gas turbines, the need for easy and reliable ignition during ground startup while accelerating to self-sustaining speed must be met [3.44]. There is also an additional requirement for rapid reignition of the chamber after flameout during flight. One of the basic requirements of a gas turbine combustion chamber is that combustion be sustained over a wide range of operating conditions. This is particularly true for the aircraft combustion room, which must sometimes operate at low temperatures and pressures and fuel/air ratios far outside the normal flammability limits for fuel/air mixtures [3.45]. Combustion must be initiated and sustained in highly turbulent air currents flowing at speeds much higher than the normal burning rate of the fuels used. Furthermore, in adverse climatic conditions where there is a risk of ingesting excessive amounts of water or ice, or during takeoff from a wet runway, the ignition system must be able to operate continuously to ensure that the engine can be reignited immediately in the event of flameout [3.44, 3.45].

Aircraft entering today's fleet have more than 80% higher fuel efficiency than the first jet aircraft of the 1950s, consuming an average of 3.5 L per passenger per 100 km [3.46]. Unlike the use of alternative fuels in ground transportation vehicles, aviation poses many more restrictions on alternative fuels due to several factors. An alternative fuel that can be used in an aircraft is defined as a fuel that can potentially mix with or replace petroleum-derived fuels and does not adversely affect engine performance, maintenance, or operational life [3.47]. First, the extreme conditions under which combustion must occur reliably and safely necessitate the use of a limited range of potential liquid fuels. Second, any proposed product must be fully interchangeable with existing jet fuel products to avoid the logistical problems of airports processing multiple fuels of different qualities and the commercial limitations that this would impose. Finally, the long life of a commercial jet requires that any candidate fuel be "backward compatible" and suitable for use with existing engine technology [3.48]. For these reasons, research on the production and use of SAFs should focus on the development of "drop-in" fuels that can be used within existing flight regulations. Determining whether the ignition and combustion characteristics of SAFs will cause problems with high-altitude reignition is very important in terms of flight safety [3.45, 3.48].

From this perspective, the design and operating system of the units that power aircraft in the aviation sector have a very complex and delicate structure. Just like a diesel or gasoline engine, a gas turbine is a type of combustion engine and operates using the cycle of intake, compression, combustion (expansion), and exhaust [3.49]. One major difference, however, is that the basic movement of a gas turbine, in contrast to the back-and-forth movement of a reciprocating engine. Gas turbines play a key role in the energy industry by providing a reliable and efficient source of power generation. They are designed to convert thermal energy into mechanical energy and eventually into electricity, making them indispensable to meet the growing demand for energy worldwide [3.50]. A gas turbine has a rotary movement, in contrast to the back-and-forth movement of a reciprocating engine. Fuel combustion is a complex process that takes place in the combustion chamber of the engine, which converts chemical energy into heat energy [3.49, 3.50]. The completeness of combustion is related to fuel properties such as atomization, vaporization, and flammability. Therefore, engine performance is affected not only by fuel performance but also by engine structure and control laws. As shown in **Figure 3.11**, the modified Brayton cycle is used in both gas turbines and jet engines [3.51].

Figure 3.11 The modified Brayton cycle for gas turbines [3.51].

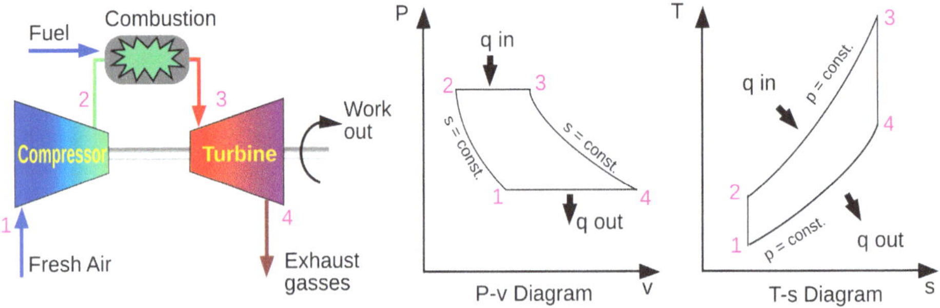

The thermodynamic process known as the Brayton cycle is fundamental to the operation of gas turbines. It similarly optimizes their performance to the Carnot cycle. This cycle maximizes the efficiency by increasing the pressure difference across the turbine, as opposed to the temperature difference in the Carnot cycle [3.52]. The power generated by gas turbines depends on multiple factors, including speed, firing temperature, and management of secondary control elements. The turbine is designed to produce a usable torque at the output shaft, while the jet engine allows most of the hot gases to expand into the atmosphere, producing usable thrust. The basic principle of a gas turbine is shown in **Figure 3.12** [3.53].

Figure 3.12 The basic principle of a gas turbine [3.53].

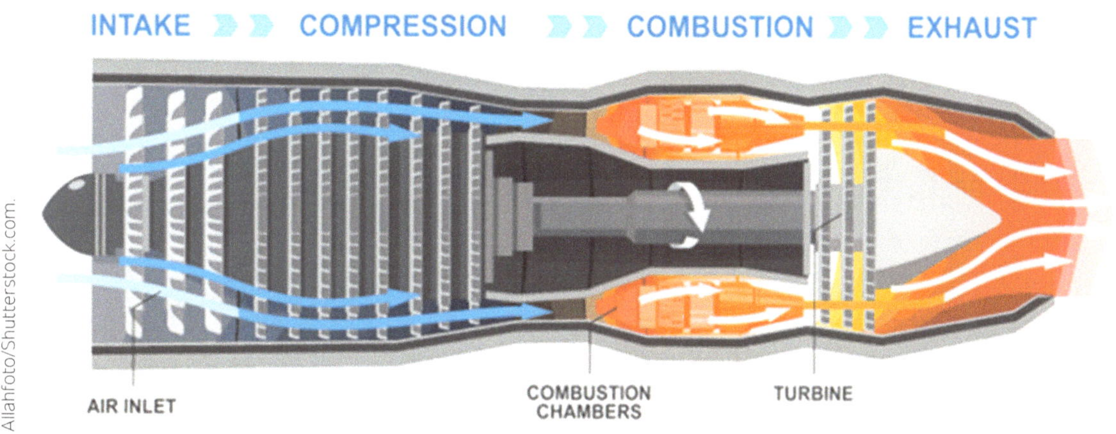

Flight safety, disconnecting the aircraft from the ground and reconnecting it with the ground after the desired distance is exceeded, requires a great deal of physical strength. Therefore, all integrated technological systems that make up these engines must work in full activation with each other. Jet engine technology is one of the most important research areas in the aviation and space industry [3.51, 3.52]. These engines, which have been used in aircraft for many years, continue to develop every day. Engine manufacturers in the aviation industry are pushing all possible conditions to produce powerful engines with low fuel requirements. There are various power units in the aviation sector with long-distance and heavy load-carrying potential. Today, apart from piston engines, there are six main categories of jet engines: turboprop engines, turboshaft engines, turbojet engines, turbofan engines, ramjet, and scramjet engines, classified according to different purposes and capacities (**Figure 3.13**) [3.54]. A jet engine draws air through air intake (or inlet) using a fan or compressor. For large passenger aircraft, the amount of air drawn in can reach over 1.3 tons/s, depending on the size of the engine.

Figure 3.13 Jet engine types [3.54].

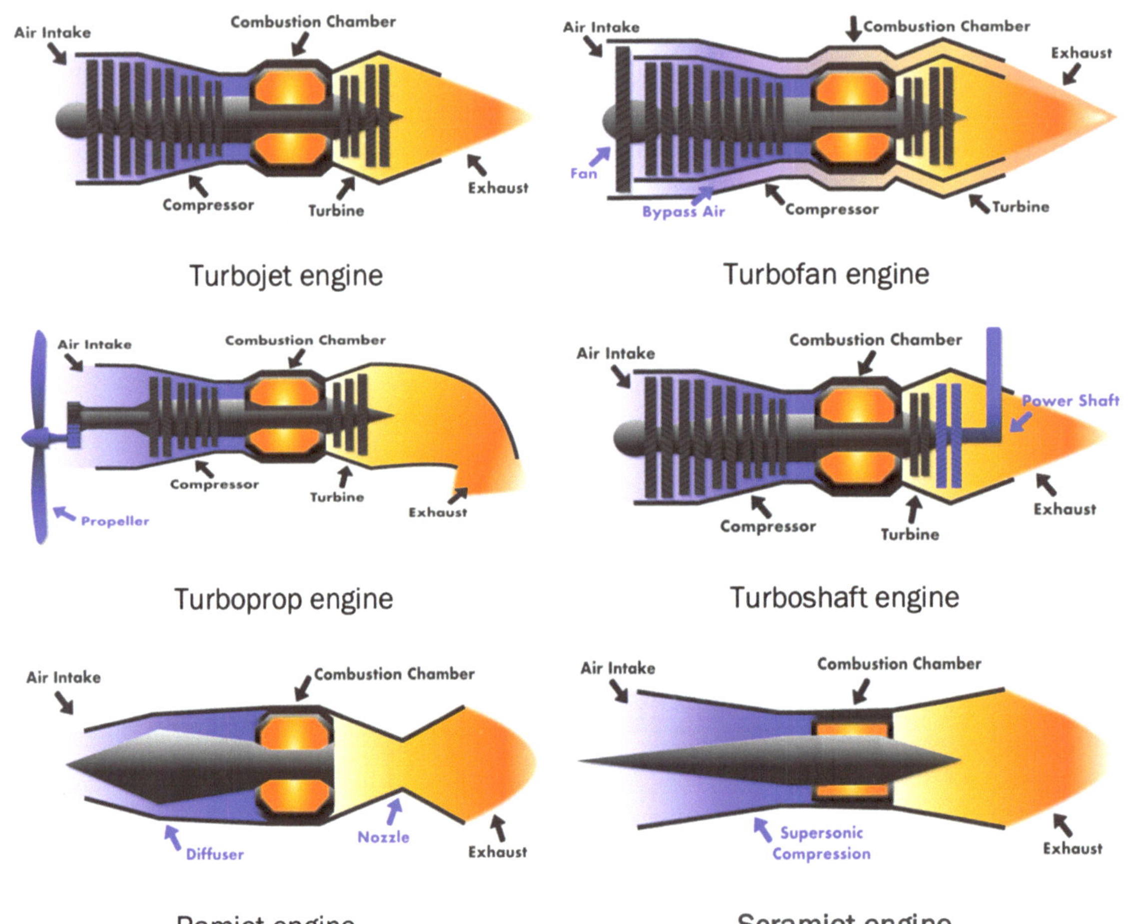

As the gas turbine increases its speed, it speeds up the compressor, pushing more air through the combustion chamber (**Figure 3.14**) [3.55]. This results in a higher combustion rate of the fuel, sending hot gases at a high pressure to the gas turbine, which in turn further increases its speed. Combustion chambers must burn stably over a wide range of operating conditions with combustion efficiency levels close to 100%. Another important requirement is that the engine burn easily and reliably during ground startup while cranking up to its self-sustaining speed. The aircraft gas turbine has the additional requirement that the combustion chamber be rapidly relit after flameout during flight. Therefore, the combustion performance parameters of primary importance to a gas turbine are combustion efficiency, stability, and ignition [3.56].

Figure 3.14 The basic principle of a gas turbine [3.55].

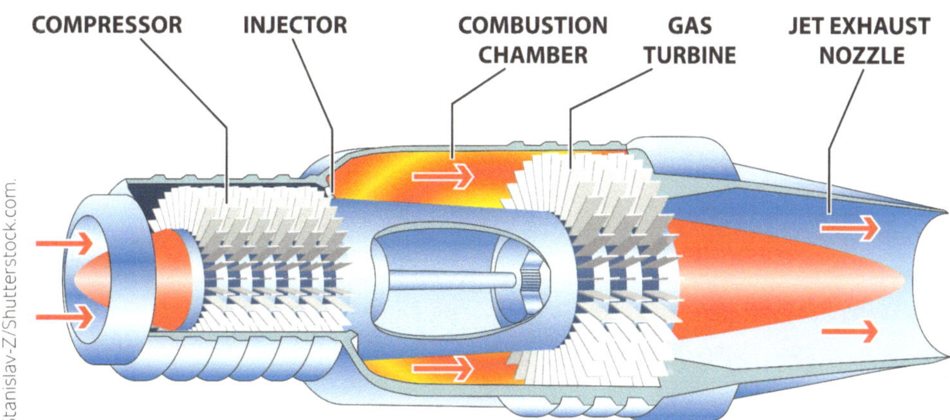

The parts of jet engines are as follows:

- **Air Intake:** This is the section where air is supplied to the engine. The main function of the inlet is to provide a continuous and smooth air supply to the compressor [3.56].
- **Compressor:** The design and shape of this section are in accordance with subsonic or supersonic flight and performance. Titanium parts called paddles, which are placed at an angle on the outer edges of the disks arranged in rows on the engine shaft, increase the speed of the air and at the same time reduce the volume of the air by performing a compression process. It sends hot air to the combustion chamber. As altitude increases, the importance of the compression ratio of air increases. Since temperature will decrease as altitude increases, compression must be sufficient. This means a higher compression ratio. This compression ratio does not pose a problem in low-altitude flights. The higher the difference between pre- and post-compressor pressure, the greater the overall efficiency and thrust [3.53, 3.56].
- **Combustion Chamber and Combustion Process:** The combustion chamber is where high-pressure air with increased temperature is mixed with the fuel. Here, a very fast air flow occurs depending on the state of the mixture. The combustion process in a gas turbine can be classified as diffusion flame combustion or lean premix staged combustion. In diffusion flame combustion, fuel/air mixing and combustion occur simultaneously in the primary combustion zone [3.57]. This creates regions of nearly stoichiometric fuel/air mixtures where temperatures are very high. In lean premix combustors, the fuel and air are thoroughly mixed in the first stage to produce a uniform, lean, unburned fuel/air mixture that is delivered to the secondary stage where the combustion reaction takes place. Manufacturers use different types of fuel/air staging, including fuel staging, air staging, or both, but the same staging, lean premix principle, applies. At the same time, since there is a very high temperature, they are made of high-temperature-resistant materials. A large amount of energy is required to ignite heterogeneous and highly turbulent mixtures

flowing at speeds of the order of 25 m/s in the combustion chamber, and the ignition of the mixture is achieved by a spark plug or arc discharge [3.58]. The ignition performance of an aircraft engine is expressed in terms of the range of flight conditions in which combustion can be restored after flameout at high altitudes. For flight safety, each ignition attempt must occur within a maximum time of 10 sec but can be as low as 3 sec [3.59].

- **Turbine:** The hot gas burned in the combustion chamber at high pressure goes to the turbine. The burned gas is transferred to the exhaust pipe at the same pressure through turbine stages [3.50, 3.57].
- **Engine Shaft:** This shaft, which is mounted along the length of the motor, houses the compressor and turbine stages. They rotate at the same speeds. In the case of single-shaft gas turbines, the fuel flow control maintains a constant speed, while twin-shaft gas turbines allow the speed to be adjusted according to the needs of the process [3.58, 3.59].

Limitations on power output at the end of the combustion reaction are generally related to the maximum ignition temperature and the maximum speed of the gas turbine. In gas turbines, failure to achieve high combustion efficiency levels is technically unacceptable. In part, combustion inefficiency represents waste of fuel, mainly because it occurs in the form of unburned HCs and pollutant emissions such as CO. Current emission regulations therefore require a combustion efficiency above 99%. For modern aircraft engines, combustion efficiency is effectively 100% at take-off conditions [3.60]. To avoid the production of "white" smoke, efficiency must exceed 96% and must not be below 90% at any point in the operating cycle. High combustion efficiency is required at this "off-design" point because the pressure and temperature of the air passing through the engine windmill and the combustion chamber are close to ambient values. At high altitudes, these are so low that stability margins are very narrow. This means that when the engine control system tries to supply more fuel to the combustion chamber to compensate for combustion inefficiency, this extra fuel can cause the flame to "go rich." Therefore, an important design requirement for an aircraft combustion chamber is that it be sized large enough to provide a sufficient level of combustion efficiency during engine restarts at the highest altitude where reignition capability is required.

When gas turbine types are evaluated according to their structure and function, turbojet and turbofan types are the most common. The turbojet engine is the most basic type of gas turbine for obtaining thrust using a jet flow and the simplest in terms of structure. It is suitable for aircraft that fly at high speeds and is used in supersonic airliners and jet fighter planes [3.61]. With the turbofan type, a separate turbine is placed after the turbojet engine, and this turbine operates a large fan at the front of the engine. This fan increases the air flow rate and boosts thrust. This type has a low noise level and low fuel consumption at subsonic speeds, so today this type of gas turbine is used almost exclusively as airliner engines. In general, the effect of fuel on the ignition process is significant in gas turbine engines but has remained constant since the commercial aviation industry uses only the specification-controlled Jet A-1. Operational dependence on Jet A-1 and the uncertainty of transitioning to an unknown product have created significant barriers to alternative HCs and/or production techniques [3.62].

In gas turbine combustion systems, fuel properties directly affect ignition performance. The most important of these effects is the density of fuel vapor concentration, especially near the spark plug and in the primary region during ignition. If aviation turbine fuel specifications are expanded to include a larger proportion of crude oil, the most significant changes will be increased aromatic content and a higher boiling point. These changes will reduce the fuel's volatility and increase its viscosity, thus deteriorating atomization quality and reducing the surface area of the spray [3.63]. Both effects will reduce the fuel evaporation rate, thus further exacerbating the combustion problem. These issues are currently not a problem for existing fuel supplies, but they will become very important if the inability of world oil production to meet basic demand necessitates using synthetic and biojet fuels derived from coal, biomass, shale, and tar sands worldwide. Increasing interest in alternative aviation fuels and developments in unconventional refining techniques enable an increased variety of SAFs. However, limited data are available on the complex relationships between these alternative fuels and gas turbine combustion performance. The current alternative fuel certification process in the aviation industry is limited to fit-for-purpose verification and requires extensive, resource-intensive, robust, full-scale testing for widespread use. Despite the development of the ASTM procedure aimed at faster alternative product approval and thus entry into service, significant uncertainties remain regarding the complex interactions between fuel chemistry and combustion process, often resulting in the need for extensive testing [3.64]. From the perspective of feedstocks and preparation processes, the biojet fuel production route is certified according to ASTM D7566. The composition of SAFs is different from those of conventional aviation fuels. For example, the maximum mass fraction of aromatics in SAFs such as FT-SPK, HEFA-SPK, SIP, AtJ-SPK, and HC-HEFAs cannot exceed 0.5%, while the sulfur content should be less than 15 ppm [3.64, 3.65]. The composition of the fuel not only is closely related to combustion performance in the combustion chamber but also affects engine exhaust emissions. This is a critical concern for engine manufacturers. The low aromatics and sulfur content of SAFs are bound to promote cleaner burning behavior in aviation engines by reducing PM and sulfur dioxide emissions. In addition to the difference in the content, SAFs and conventional aviation fuels exhibit some differences in chemical properties. According to ASTM D7566, the final boiling points of Jet A/Jet A-1 fuel and SIP fuel are limited to 300 and 255°C, respectively. The viscosity ranges at 15°C for Jet A/Jet A-1 fuel and FT-SPK fuel are 775–840 and 730–770 kg/m^3, respectively [3.65, 3.66]. Differences in volatility, thermal stability, and viscosity cause SAFs to exhibit different atomization and combustion performance compared to conventional aviation fuels in current aircraft engines. This may affect pollutant emissions such as CO, NO_x, and unburned hydrocarbons (UHCs) to some extent, but engine operating conditions have a greater impact.

Current ASTM certification standards for blending of alternative fuels with petroleum-derived jet fuels restrict the blend ratio to be no more than 50/50 by volume [3.67]. In other words, despite the qualitative predictability from experimental large-scale test results and the usefulness of such tools, there does not appear to be a quantitatively predictable relationship between the physical and chemical kinetic properties of jet fuels and their global combustion behavior. The development of SAFs is currently receiving increased attention. Using a range payload approach, the need for drop-in fuels is justified. SAFs are derived from renewable resources and are designed to be drop-in replacements for conventional jet fuels, meaning that they can be used in the existing aircraft engines and infrastructure without modifications. When blending jet fuel with SAFs, determining the density, aromatic content, and GHG footprint of the blended fuel is important for flight safety. Currently, little is known about their fit-for-purpose properties (lubricity, dielectric strength, octane number), combustion operability and emissions, material compatibility, and blending issues [3.68].

Available alternative fuels can be categorized into two types depending on whether the product increases supply security or provides a reduced environmental footprint. As an alternative fuel in gas turbines, **hydrogen** is the closest source to an ideal fuel in terms of combustion performance [3.69]. High flame speeds, wide combustion limits, easy ignition, and lack of soot formation are some of the characteristics of hydrogen. Moreover, liquid hydrogen has a cooling capacity that is far superior to all other fuels. The main disadvantages of hydrogen are its very low density and low boiling point, which necessitate the use of large, highly insulated storage tanks in aircraft. It is also very costly to produce. For these reasons, extensive research needs to be carried out on the use of hydrogen with respect to flight safety [3.69, 3.70].

Alcohols are not practical to be used as fuels in long-range aircraft because of their water retention at low temperatures, high oxygen content, and consequently low calorific value [3.71]. This prevents cold starts, part loads, and transient operation at low vapor pressure. Lighter alcohols (C1-OH and C2-OH) are considered safer than kerosene because of their higher flash points and because fires from these alcohols can be extinguished with water. Alcohols burn with a low-luminosity blue flame and minimal exhaust smoke and have wide flammability limits [3.72]. Additionally, the low flame temperature results in relatively low NO_x emissions. However, they tend to be corrosive to some metals, and special precautions are required to avoid this problem. As a result of the combustion of alcohols, aldehyde emissions are released, which cause ozone pollution. It is essential that they are used in mixtures to improve combustion properties.

Synthetic fuels stand out in terms of combustion performance compared to hydrogen and alcohol fuels. Significant work should be performed to assess the GHG footprint of FT, GtL, and conventional fuels. The main benefits of FT transportation fuels are huge and secure local supply, which makes them clean-burning fuels (very low in nitrogen, aromatics, and sulfur) [3.73]. A limited number of studies, including a study by Rye et al., compared the combustion performance of GtL, Jet A-1, thermally stressed Jet A-1, and diesel fuel, concluding that changes in fuel chemistry (including significant changes in HC composition) do not affect ignition performance provided that sufficient light HCs are present in the primary region of the combustion chamber [3.74]. Research on gas turbine durability, atomization, and ignition tests has shown that there is no significant difference between the performance limits of fully synthetic jet fuels and Jet A-1 fuels.

Biofuels that can be produced from biomass are potential new players among aviation fuels. Fuel derived from biomass has little or no sulfur or ash content and has a net-zero CO_2 footprint as CO_2 is consumed in the production of renewable biomass [3.75].

In a study by Won et al., the prevaporized spherical combustion behaviors of a petroleum-derived jet fuel (JP-8), five alternative jet fuels (Shell Synthetic FT-SPK, Sasol CtL-IPK, HEFA-HRJ Camelina, HRJ Tallow, and Gevo AtJ), and five 50/50 (by volume) JP-8/alternative fuel blends were experimentally investigated and compared [3.76]. Three experiments were conducted to investigate the gas-phase combustion behavior of the tested fuel samples: (1) global oxidative species profiles in a variable-pressure flow reactor, (2) diffusion flame extinction in a counterflow burner, and (3) premixed flame initiation in a heated global combustion chamber. Laminar flame speeds of all measured fuels showed little variation comparable to the experimental uncertainty, and combustion performance efficiencies were found to be adequate. As a result, Shell SPK and Gevo AtJ were found to have distinct combustion behaviors for both low-temperature reactivity and diffusive extinction.

Gas turbines typically operate at high loads (greater than or equal to 80% of rated capacity) to achieve the maximum thermal efficiency and the highest combustion chamber flame temperature [3.77]. The combustion properties of highest interest are those that govern the flame temperature, the rate of chemical reaction, the burning range, and the tendency to form soot. At reduced loads (less than 80%) or during frequent load changes, combustion chamber flame temperatures are expected to be lower than at high load temperatures, resulting in lower thermal efficiencies and incomplete combustion [3.63, 3.77].

SAFs generally exhibit favorable combustion properties, which include the following:

1. High cetane number: This ensures better ignition quality and smoother engine operation; the low aromatic content results in cleaner combustion, reducing the formation of soot and PM.
2. Cold flow properties: SAFs need to perform reliably under various temperature conditions, especially at high altitudes where temperatures can drop significantly [3.78]. The cold flow properties of SAFs, such as freeze point and viscosity, are critical performance metrics. HEFA and FT-SPK both typically have good cold flow properties, with freeze points well within the acceptable range for aviation use (e.g., below −40°C) [3.79]. This ensures that the fuel remains fluid and can be efficiently pumped and burned in the engine.
3. Energy density and efficiency: SAFs are designed to have energy densities comparable to those of conventional jet fuels, which is crucial for maintaining aircraft performance in terms of range and fuel efficiency [3.76]. For instance, HEFA-based SAFs typically have an energy density of around 43–45 MJ/kg, similar to that of Jet-A fuel (42.8 MJ/kg). This ensures that the aircraft can operate efficiently without sacrificing performance. FT-SPK also exhibits high energy density, comparable to that of conventional fuels, making it suitable for long-haul flights [3.76, 3.78].
4. Sustainability: The use of SAFs should be tested under real flight conditions, and suitability of the fuel should be determined [3.79]. Flight trials include the evaluation of engine performance during all phases of the flight, as shown in **Figure 3.15** [3.80]. These are the basic real test criteria to determine SAF performance, including a series of extreme "maneuvers" (e.g., shutting down the engine in flight and ensuring that it can be restarted). The flight should be planned to simulate real flight conditions rather than a laboratory environment, and the combustion performance of the SAF should be tested under real conditions for widespread use.

Figure 3.15 An example of a flight profile for SAF trials [3.80].

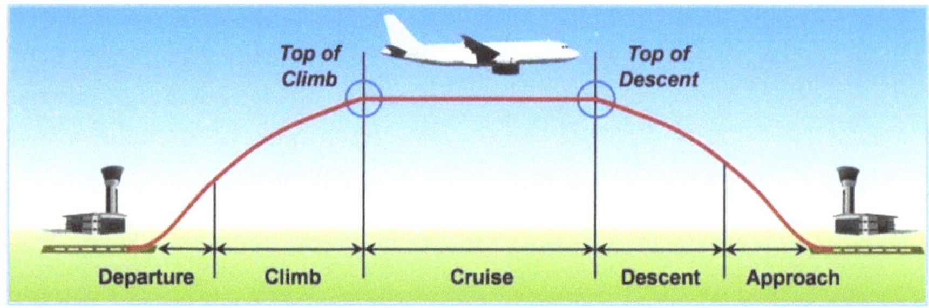

Emission Characteristics of SAFs

Commercial aviation's climate change impact is complex, reflecting the variety of emissions from operations at the surface up to cruising altitudes as high as 43,000 ft, across continents and oceans, and over varied time spans [3.81]. Emission levels are generally high, but vary depending on engine conditions, fuel type, and operating modes such as idling, taxiing, takeoff, climbing, and landing. The concentration levels of pollutants in gas turbine exhausts can be related directly to the temperature, time, and concentration histories of the combustion process [3.57]. As shown in **Figures 3.16** and **3.17**, approximately 90% of aircraft emissions occur 3000 ft above the ground, while the remaining 10% are emitted during taxiing, takeoff, initial climb, and approach and landing [defined as the landing–takeoff (LTO) cycle] [3.82, 3.83].

Figure 3.16 Aviation emissions by altitude [3.82].

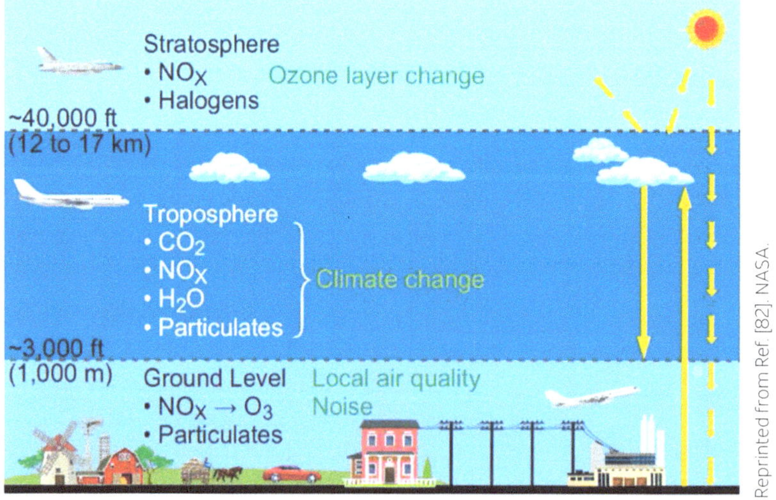

Figure 3.17 The basic definition of aircraft emissions by LTO [3.82, 3.83].

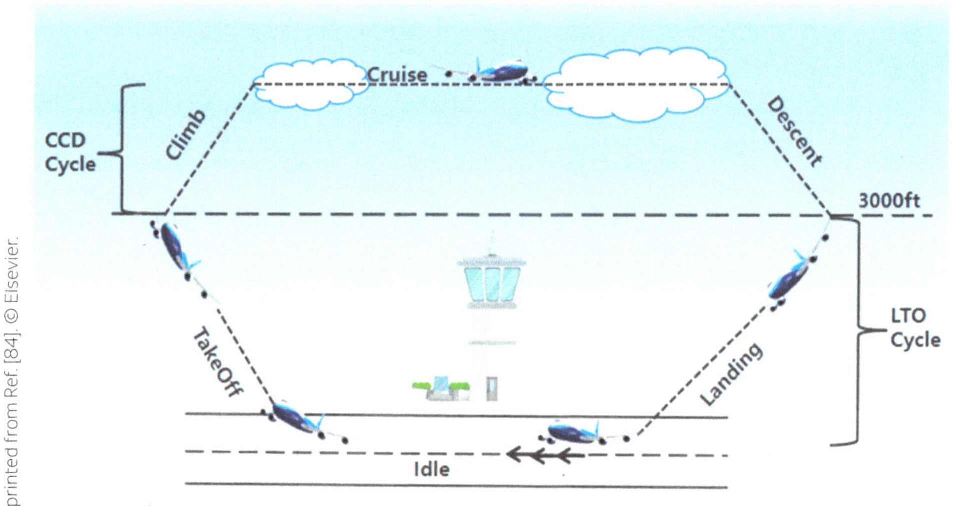

The exhaust from an aircraft gas turbine is composed of CO, CO_2, water vapor (H_2O), UHC, PM (mainly carbon), NO_x, and excess atmospheric oxygen and nitrogen [3.83]. Aviation PM emissions can be categorized as primary and secondary PM. Primary PM refers to solid and liquid particles emitted directly from the engine without undergoing chemical transformations in the atmosphere. Secondary PM refers to nvPM, and volatile particulate matter (vPM) constitutes the primary PM emissions from aviation activities [3.76]. These vary from one gas turbine to another and with changes in operating conditions for any combustor. The nature of pollutant formation is such that CO and UHC concentrations are highest at low power conditions and decrease with an increase in power. In contrast, NO_x and smoke are relatively insignificant at low power settings and reach maximum values at the highest power condition [3.78, 3.81].

The current subsonic fleet largely use fossil-based kerosene as an energy source, which, when combusted in the aircraft engine, results in the emission of a number of products of complete and incomplete combustion. The climate impacts of aircraft emissions are depicted in **Figure 3.18** [3.84]. The emitted emissions can have both warming and cooling effects. The effects of non-CO_2 gases depend on weather, flight path, time of the day, fuel characteristics, and engines. Persistent contrails and NO_x emissions have the greatest impact.

Figure 3.18 The climate impacts of aircraft emissions [3.84].

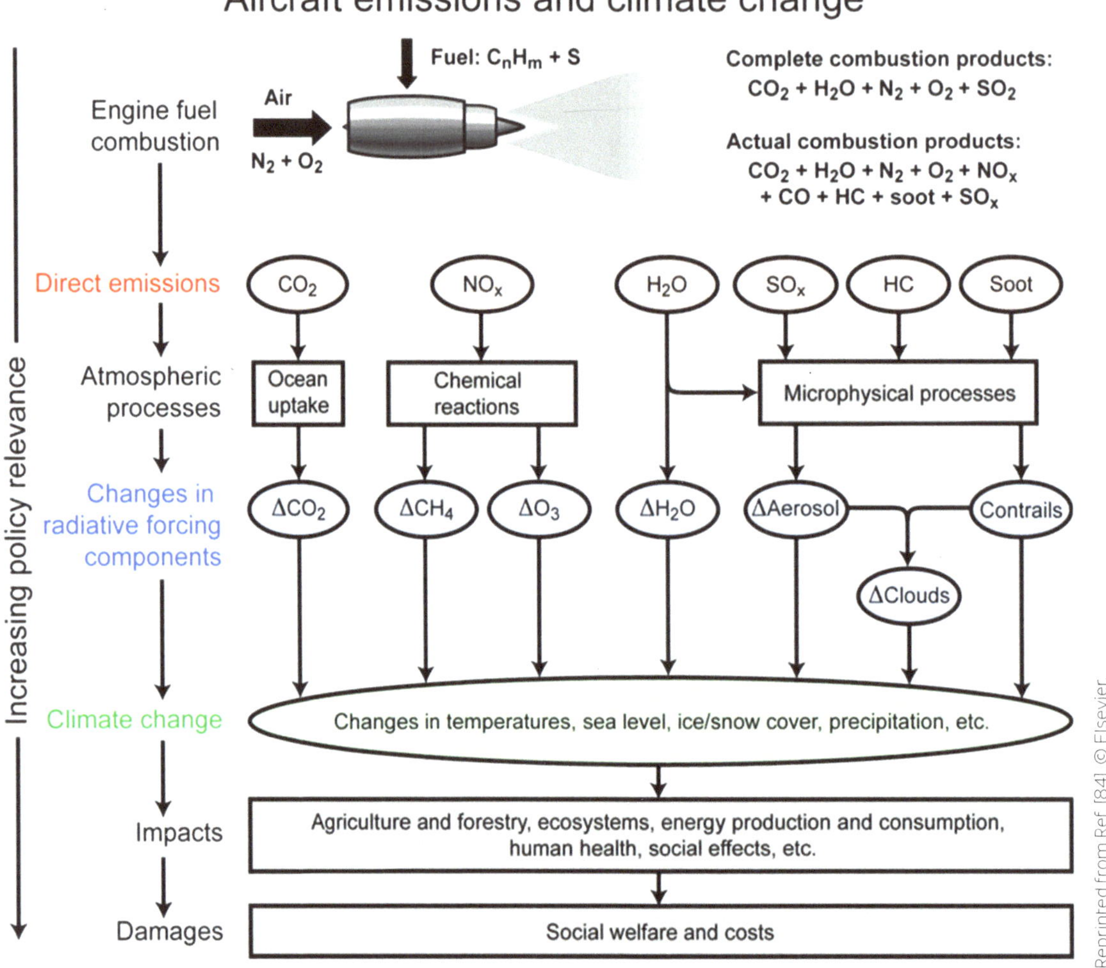

In line with its fuel consumption, aviation accounts for 2–3% of global GHG emissions. The US, which has the world's largest commercial air traffic system, also plays a leading role in aircraft-related emissions [3.84]. EPA reports that commercial airplanes and large business jets contribute 10% of US transportation emissions and account for 3% of the country's total GHG emissions. Non-CO_2 emissions both warm and cool the atmosphere and contribute to additional global warming overall. The largest component of aircraft emissions is CO_2, accounting for approximately 70% of emissions [3.85]. CO_2 emissions of a flight can be calculated directly from fuel consumption. Jet fuel consumption produces CO_2 at a defined rate (3.16 kg of CO_2 per kilogram of fuel consumed) regardless of the phase of flight. One ton of CO_2 from an aircraft engine has the same effect as 1 ton of CO_2 from another source such as heating or car traffic [3.81, 3.86]. Once released, 30% of a given amount of the gas is naturally removed from the atmosphere within 30 years, an additional 50% disappears within a few hundred years, and the remaining 20% has the potential to remain in the atmosphere for thousands of years. The full impact of warming due to aviation includes both CO_2 and non-CO_2 effects, with contrail-induced warming likely being the most significant driver of non-CO_2 effects of aviation [3.87]. When aircraft fly at high altitudes and discharge black carbon particles, they create condensation trails. These contrails can form cirrus clouds, which have warming and cooling effects. As shown in **Figure 3.19**, contrails are clouds formed when H_2O condenses and freezes around small particles (aerosols) in aircraft exhaust [3.88].

Contrails created by aircraft flying in cold, humid air can significantly warm the climate. Most contrails dissolve within a few minutes, but under certain conditions, they can persist in the atmosphere, spread out, and form artificial cirrus clouds that trap the heat that escapes [3.87, 3.88].

These clouds can last for more than half a day and, under certain atmospheric conditions, coalesce and spread over thousands of square miles, spreading the heat-trapping effect over large areas. For most contrails, the heating effect dominates during the day, except for optically very thick condensation traces that block most of the incoming sunlight [3.89].

During the night, there is no solar radiation, so the cooling effect disappears, and nighttime condensation traces become on average warmer than daytime contrails. The effect of contrail cirrus clouds on global warming is stronger in North America and Europe, the regions with the heaviest air traffic [3.90]. However, as air travel increases, it is also expected to increase significantly in Asia. Recent forecast models predict that the global heat trapping effect of contrails will triple by 2050 unless airlines and aircraft manufacturers significantly reduce emissions or air traffic patterns change. This represents an additional climate impact on top of direct emissions from aviation.

All the remaining emissions shown in **Figure 3.18** make up less than 1% of the exhaust plume. Measuring the amount of non-CO_2 emissions that aircraft emit at high altitudes while flying is still difficult [3.81, 3.84]. NO_x in the exhaust chemically forms ozone (O_3), creating a warming effect, but it also eliminates methane (CH_4), a powerful GHG that creates a cooling effect as it decreases in the atmosphere [3.82]. Other emissions include HCs, soot, and sulfates. Sulfates reflect sunlight, providing a small cooling effect. Soot absorbs heat, and these black carbon particles readily fuse into ice crystal nuclei. Modern jet engines emit far fewer soot particles than earlier models, reducing their contribution to condensate trails and eliminating the black exhaust that was typical of jet aircraft decades ago [3.50, 3.55]. This is mainly because these emissions have both warming and cooling effects, depending largely on the state and composition of the surrounding atmosphere.

Figure 3.19 Contrail formation and climate effects [3.88].

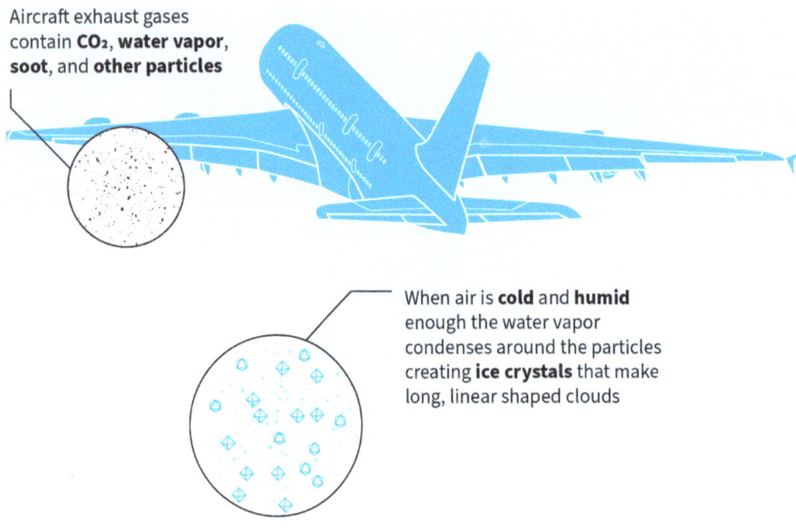

If the aircraft flies through ice super-saturated (ISSRs) regions the ice crystals in the contrails can persist, spread over time and create cirrus clouds that can linger in the atmosphere

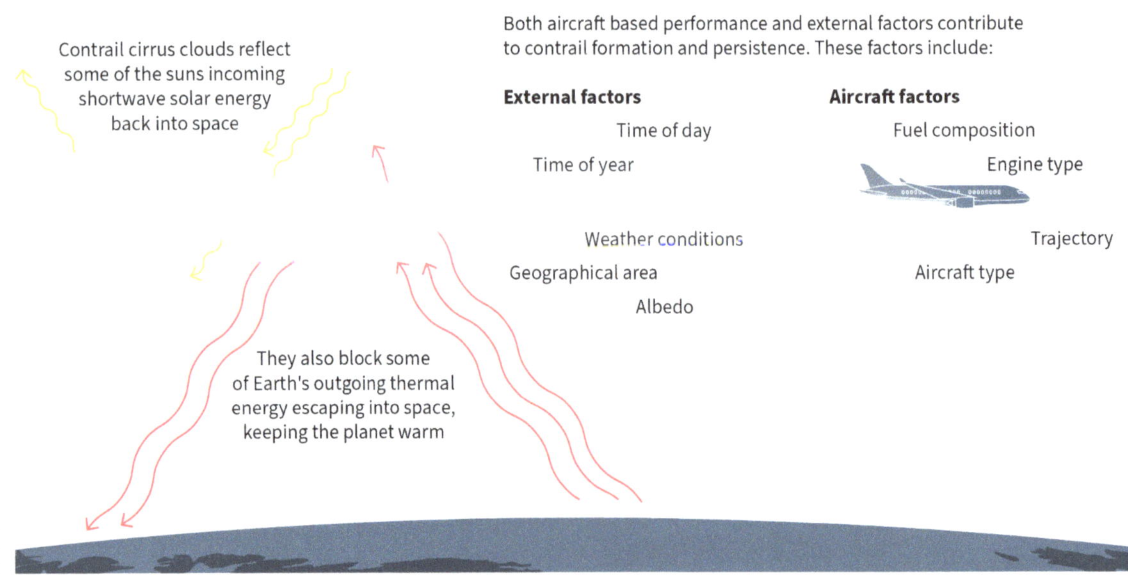

Depending on operating conditions and fuel composition, typical exhaust emissions from a stationary gas turbine fall into two distinct categories: major and minor pollutants [3.91]. Measurements of exhaust gases from jet engines must be made using ground test stands, but emission levels change when the engine is flying at high altitudes. Non-CO_2 emissions are not proportional to fuel consumption. For this reason, today's aircraft engines can show great differences in their emissions.

The major species (CO_2, N_2, H_2O, and O_2) are present in percent concentrations. Minor species (or pollutants) such as CO, UHC, NO_x, SO_X, and particulates (PM) are present in parts per million concentrations [3.92]. The formation of NO_x is highly dependent on the high temperatures generated in the combustion chamber. CO, volatile organic compounds (VOCs), and PM are primarily the result of incomplete combustion. Trace or low levels of sulfur dioxide (SO_2) are emitted from gas turbines. Ash and metallic additives in the fuel can also contribute to PM in the exhaust. SO_X are produced only in significant amounts if heavy oils are burned in the turbine. Emissions of sulfur compounds, primarily SO_2, are directly related to the sulfur content of the fuel. In general, given the fuel composition and engine operating conditions, the major species compositions can be calculated [3.93]. Minor species cannot be calculated, except for total SO_X. Characterization of pollutants requires careful measurement and semitheoretical analysis. Their general characteristics and impacts are presented in **Table 3.1** [3.94].

Table 3.1 General characteristics and effects of CO_2 and non-CO_2 emissions [3.94].

	CO_2 emissions	**Non-CO_2 emissions**
Characteristics	CO_2 is a greenhouse gas	Other main climate-impacting emissions from aviation are not greenhouse gases
	The amount depends on the type of fuel and is proportional to consumption	The amount is not proportional to CO_2 and depends on the engine and environment
		Contrails: almost entirely produced from water vapor in the atmosphere
Impact	Always warming	Warming and cooling
		Persistent contrails strongly warming at night, cooling during the day
	Can be calculated accurately	Cannot be calculated accurately
	Once emitted, impact continues for hundreds of years	Impact lasts minutes or hours
		For nitrogen oxides, a maximum of several decades
	Impact increases even when emissions remain the same	Impact does not increase although emissions remain the same
	Impact does not depend on the location, time of day, or weather conditions	Impact depends on the location, time of day, and weather conditions

Pollutant emissions from combustion processes have become a major concern for the public because of their impact on health and the environment. However, emission reduction regulations, such as those in ground transportation, have not been able to be firmly regulated in the aviation sector due to acceptable safety concerns [3.95]. Environmental organizations now recognize that air transportation is one of the fastest-growing energy-consuming sectors in the world in terms of fuel consumption. The worldwide interest in regulating and reducing gas turbine emissions and the enactment of federal and state regulations in the US have necessitated research on gas turbine exhaust emission predictions and the effect of exhaust emission control methods on gas turbine performance. In recent years, there have been rapid changes in both the regulations for controlling gas turbine emissions and the technologies used to meet these regulations. Emissions from both turbines and jets are similar, as are their control methods [3.96]. The emissions are primarily UHCs, unburned carbon that results in the visible exhaust, and NO_x. Control of UHCs and unburned carbon may be accomplished by redesigning fuel spray nozzles and reducing cooling air to combustion chambers to permit more complete combustion. Although the formation of CO reduces CO_2 emissions, the amount of CO is insignificant compared to the amount of CO_2 produced. The fact that the majority of fuel carbon is not converted to CO_2 is due to incomplete combustion [3.97]. US airlines have converted their jet fleet to lower-emission engines using these control methods. NO_x emissions may be minimized by the reduction in maximum temperature in the primary zone of combustors [3.91, 3.96]. Other climate impacts (non-CO_2 emissions and warming effects) need to be decreased or offset using other measures. SAFs are central to these targets. The increased use of SAFs will reduce fossil CO_2 emissions and the climate impact of non-CO_2 emissions in the future. Exhaust emissions of concern and emission control techniques can be divided into several categories, as shown in **Table 3.2** [3.91]. There are three generic types of emission controls in use for gas turbines: wet controls using steam or water injection to reduce combustion temperatures for NO_x control; dry controls using advanced combustor design to suppress NO_x formation and/or promote CO burnout; and post-combustion catalytic control to selectively reduce NO_x and/or oxidize CO emission from the turbine [3.96, 3.97]. Other recently developed technologies promise significantly lower levels of NO_x and CO emissions from gas turbines with the diffusion combustion type.

Table 3.2 Jet engine emission control techniques [3.91].

Pollutant	Control techniques
NO_x	Wet control (water/steam injection) and dry NOx emission reduction (fuel-lean injection)
CO	Combustor design catalytic reduction
UHC and VOC	Combustor design
SO_x	Control sulfur in fuel
Particulates and PM-10	Fuel composition
Smoke reduction	Combustor design–fuel composition–air atomization
Particulate reduction	Fuel composition–sulfur–ash

The increasing need to conserve limited fossil fuel resources can only be met by increasing engine cycle efficiency, which will reduce fuel consumption. In practice, this has traditionally required the modification of the engine combustion process. This approach reduces CO_2 emissions but results in higher combustion temperatures and higher NO_x levels. The desire to burn less fuel and therefore produce less CO_2 is in direct conflict with the equally important need to reduce NO_x [3.98]. Therefore, strategies to reduce emissions, including changing the fuel source, have been developed. Aviation emissions, particle size, and emission levels depend on aircraft type, engine conditions, fuel type, and operating modes. On the other hand, airports are significant sources of high emissions, and human exposure to these emissions is becoming a growing public health concern. When pollutant emissions are evaluated in the aviation sector, a multifaceted strategy is necessary, as shown in **Figure 3.20** [3.99]. The total emission components emitted in the aviation sector directly affect all individuals, including passengers staying in that area and workers. Aircraft emissions cause lung and systemic inflammation, which can potentially lead to cancer, asthma, and respiratory and coronary heart diseases [3.100]. The largest part of these emissions is pollutants formed as a result of the combustion from the engine. The reported adverse health effects of jet engine emissions are similar to those caused by exposure to diesel exhaust and air pollution. However, given the lack of consensus and sensitivity on optimum measurement methods, equipment, and quality control for near- and far-field airport emissions and human risk assessment markers, further studies on exposure duration and toxicological mechanisms are needed [3.99, 3.100]. It should be noted that PM, especially polycyclic aromatic hydrocarbons (PAHs) and other particulate-bound organic compounds, is more important than particle mass in determining the effects of air pollution [3.101]. In this context, time- and size-resolved PM measurements are necessary to test mechanistic toxicological hypotheses, characterize the relationship between combustion operating conditions and transient emissions, and develop air quality plans [3.102].

Figure 3.20 Aviation sector emission control strategy [3.99].

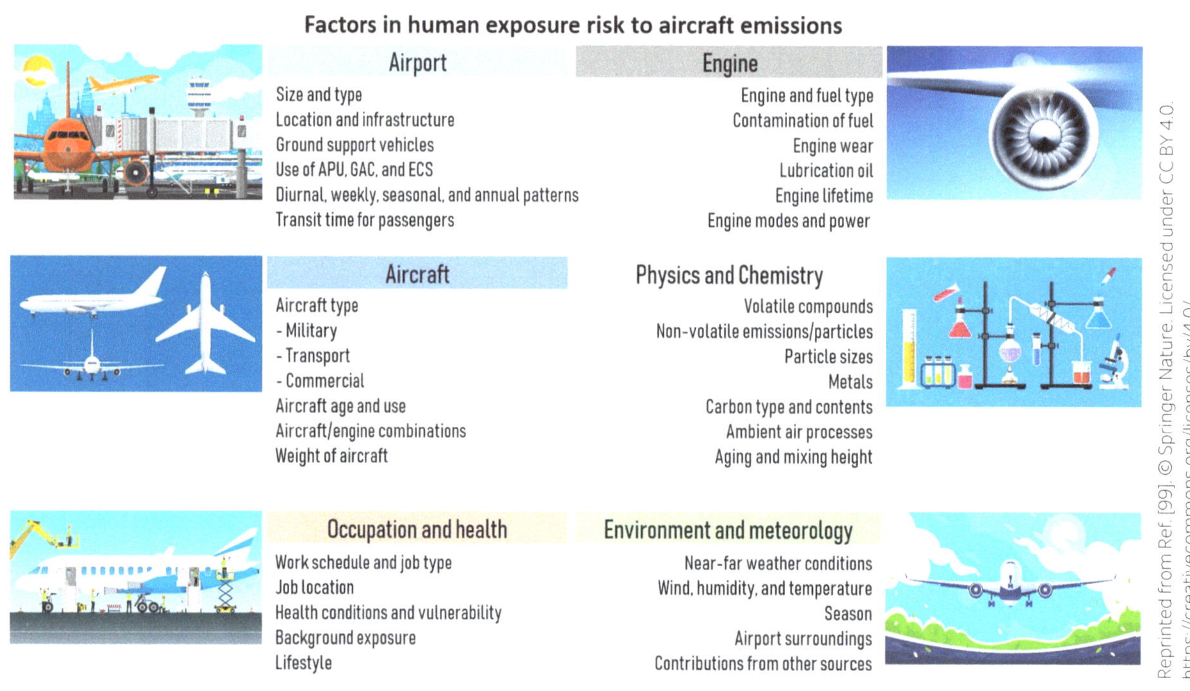

Strategies developed to reduce pollutants originating from gas turbines include the following:

- **Adjustments to aircraft technology:** Aviation emissions from international flights have not been included in the international climate regime administered by the United Nations Framework Convention on Climate Change (UNFCCC), as they fall outside of the scope of nationally determined climate action. Instead, these emissions have been dealt with by the ICAO [3.103]. CORSIA is a global offset scheme under which airlines and other aircraft operators will offset any increase in CO_2 emissions above the 2020 levels. This states that aviation's net CO_2 emissions will be fixed, and other emissions reduction measures such as technology, SAF, operations, and infrastructure options should be pursued. In 2016, the ICAO established two-tiered CO_2 emission standards for new aircraft [3.104]. In addition, aircraft engine certification requirements take into account CO, HCs, NO_x, and smoke emissions in the control mechanism. Improvements in engine thermal efficiency not only reduce direct operating costs but also reduce pollution. Unlike ground vehicles, which do not need to be fuel efficient because they can refuel frequently, long-haul aircraft must carry all their fuel on board. It is expensive and heavy and takes up significant storage space. Its weight limits the aircraft's range, and it must be stored in tanks, which affects wing size and reduces maximum payload. The ICAO has implemented regulations for civil subsonic turbojet/turbofan engines with nominal thrust levels above 26.7 kN for a defined LTO based on an operational cycle around airports [3.103, 3.104, 3.105, 3.106]. A more restrictive efficiency standard applies to designs certified after January 1, 2020, for commercial jets and after January 1, 2023, for business jets. Each aircraft category enters service approximately four years after certification. The efficiency requirements will apply to all new aircraft deliveries starting from January 1, 2028. The standards are based on the mass of an aircraft and will require, on average, a 4% reduction in cruise fuel consumption compared to the performance of new aircraft delivered in 2015 [3.104]. In the near future, combustion performance and emission reduction will become more prominent in aircraft technology.

According to NASA's technological projection for gas turbines, shown in **Figure 3.21**, the prominent approaches to combustion improvement and emission reduction are specified [3.82]. According to this projection, by making SAF production more efficient and expanding its use, its integration into existing gas turbine fuel systems and emission reduction potential will emerge. To ensure that the emission reduction potential of SAFs is maximized, they must be usable without conventional kerosene. Current SAF pathways are limited to 50% blend proportions due to a lack of chemical equivalence of traditional fuels and SAFs [3.64, 3.65, 3.66, 3.67]. This arises as most current SAF pathways only replicate the paraffin content of jet fuel, with limited options for including other components. These most notable aromatics are necessary for ensuring engine longevity [3.107]. While new SAF pathways may allow for 100% SAF utilization in the future, engine design changes will likely also be required to maintain optimal engine performance, given the different chemical compositions (particularly affecting combustion temperature) of SAFs versus conventional kerosene. In this context, CORSIA is expected to continue to play an important role in the expansion of international carbon markets in the future.

Figure 3.21 Challenges for aviation aircraft technology [3.82].

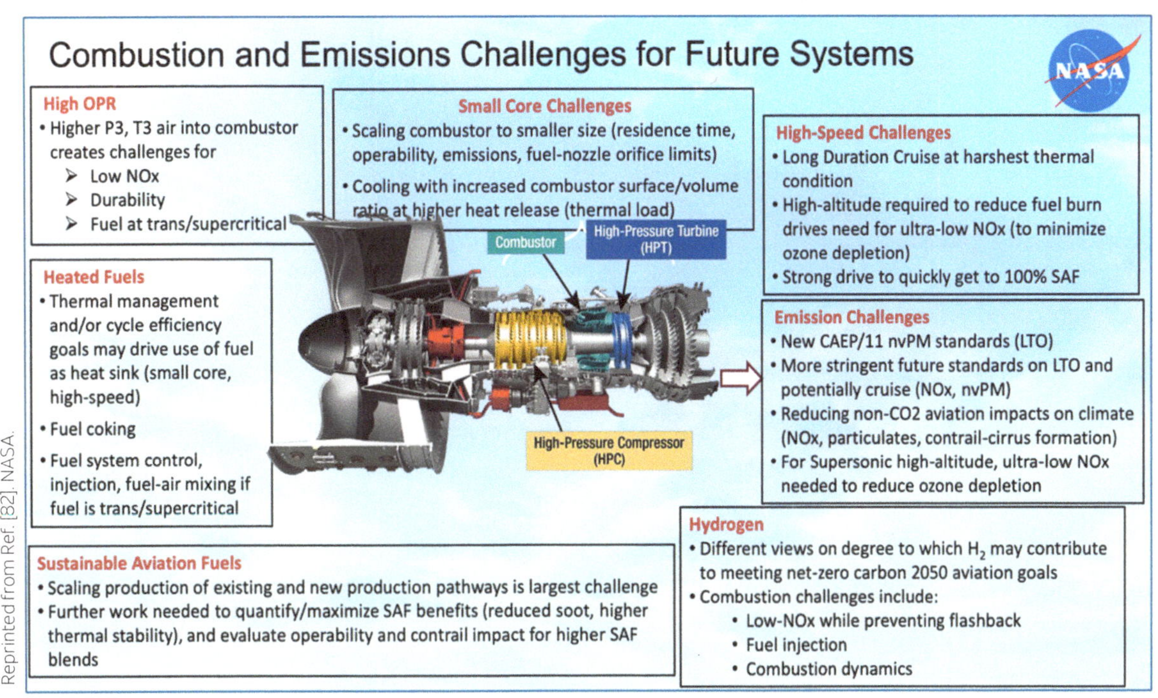

- **Adaptation of flight operations:** If the formation of persistent contrails at night could be avoided, clouds formed by aircraft could have an overall cooling effect [3.108]. The zones that enable the formation of strongly warming contrails are often only a few hundred meters thick and limited in size. Fuel efficiency is critical to the future of aviation, not just for environmental reasons but also for financial reasons. Fuel accounts for more than 30% of airline operating costs. The most effective way to avoid these costs is to adjust flight altitude, with the caveat that fuel consumption increases when an aircraft deviates from its ideal altitude or its flight path, which also increases CO_2 emissions [3.109]. The crucial point is having exact, real-time temperature and humidity information for the relevant altitudes (usually over 8000 m).
- **Adjustments to fuels:** Fuel selection for a gas turbine is determined by cost, availability, and ease of use. It also depends on the significant and limiting effects of aircraft engine requirements on the physical and chemical properties of the fuel [3.110]. It is common to classify the HCs found in fossil-derived gas turbine fuel into four main groups: paraffinic, olefinic, naphthenic, and aromatic. Understanding how different fuel components affect pollutant emissions and the resulting impacts on human health and the environment is important for researchers and policymakers. The proportions of these groups largely determine the character of the fuel. A typical Jet-A fuel composition breakdown is shown in **Figure 3.22** [3.111].

Figure 3.22 Jet-A fuel composition [3.85].

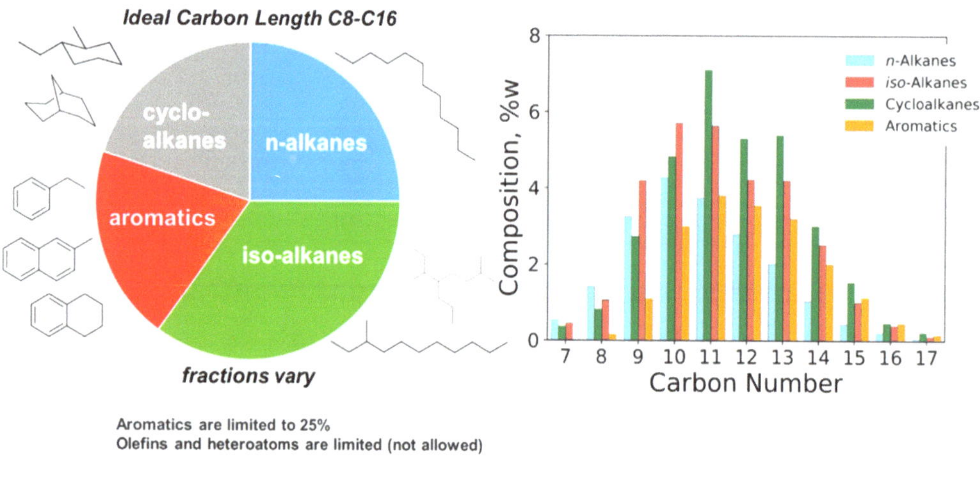

Alkanes have a chemical formula of C_nH2_{n+2}. In organic chemistry, an alkane, or paraffin (-cyclo), is an acyclic saturated HC. Methane (CH_4), ethane (C_2H6), propane (C_3H_8), and butane (C_4H_{10}) are the first four alkanes. Current aviation fuels contain an average of 80% alkanes depending on the crude oil source and distillation process. Paraffins tend to have a higher hydrogen/carbon ratio, lower density, lower freezing point, and higher gravimetric heating value than other types of HC fuels [3.111, 3.112]. They have high thermal stability, and their combustion is characterized by coke accumulation and freedom from exhaust fumes [3.113].

Olefins have a chemical formula of C_nH_{2n}. They are not normally found in crude oil but are produced by conversion processes in the refinery. Because olefins are unsaturated, meaning their molecules contain fewer than the maximum possible hydrogen atoms, they are chemically active and react readily with many compounds to form resinous gums and rubber-like materials [3.114]. For this reason, olefins are very undesirable in gas turbine fuels and are only found in trace amounts [3.113, 3.115].

Naphthenes have a chemical structure of CH_{2n}. They are saturated HCs in which carbon atoms are bonded to form rings instead of chains, as in paraffins [3.114, 3.115]. They exhibit similar properties to paraffins in terms of their chemical stability, high gravimetric heat of combustion, and low soot formation tendency.

Aromatic compounds are ring compounds containing one or more six-membered rings equivalent to three double bonds. The chemical formula of aromatics is C_nH_{2n-6}. Jet fuel contains aromatics because crude oil from which it is produced contains aromatics. Although aromatics are undesirable in gas turbine fuels, current practice in aviation fuels is to limit their aromatic content to no more than 25% by volume [3.116]. Aromatic compounds in fuels cause the swelling of the o-ring, and this helps seal the high-pressure aircraft fuel system. Disadvantages of aromatic compounds include a marked tendency to form soot and high hygroscopicity, which can lead to the precipitation of ice crystals when the fuel is exposed to low temperatures [3.116, 3.117]. Although similar in structure to naphthenes, they contain less hydrogen, and as a consequence, their specific energy is appreciably lower. Aromatic components (including benzenes, naphthalenes, and other monoaromatics), while important for engine operation, have been shown to contribute considerably to nvPM emissions (i.e., soot). Furthermore, recent emission measurements have shown that the removal of naphthalenes from jet fuel, while keeping the total aromatic content unchanged, can dramatically reduce the emissions of soot [3.118].

Given the strong current and future limitations on fuel efficiency and emissions, alternative fuels will be an energy source that needs to be researched and produced for aircraft. When an engine is certified, it must meet the emission standards recommended by ICAO/Committee on Aviation Environmental Protection (CAEP) [3.119]. Similarly, IATA is increasingly supporting technological developments in SAFs and their practical application on a global scale, aiming to achieve net-zero emissions in the aviation sector by 2050.

The use of SAFs has the benefit of reducing both fossil CO_2 emissions and non-CO_2 emissions. SAFs produce fewer black carbon particles and less sulfur dioxide. Contrails become more permeable to heat radiation when SAFs are used, which reduces the warming component of contrails [3.120]. Current research on SAF emissions includes limited research on emissions performance due to high fuel consumption and costs during engine testing, and studies on pollutant emissions are rare. Studies examining aircraft emissions and climate impacts suggest that more research is needed to fully understand the formation and impact of condensate trails and condensate-induced clouds and to develop mitigation strategies.

One of the most effective mitigation strategies is to use sustainable biofuels blended with kerosene jet fuel, which has entered the commercial aviation market. Biofuels have a significantly lower lifecycle GHG assessment than traditional petroleum-based jet fuel. SAF blends also have the potential to reduce exhaust soot, H_2O, and sulfates [3.121]. In terms of aromatic content, one way to reduce the average aromatic concentration in the fuel blend is to blend SAFs in increasing volumes without increasing the aromatic concentration in the fossil portion of the blend. Since reducing aromatic concentration in the refinery is costly, there is a possibility that refiners may blend aromatics in jet fuel to the applicable limit to reduce costs and thereby eliminate the non-CO_2 climate impacts of blending SAFs. Fewer particulates and less water vapor mean a reduction in condensate trail formation. Reducing sulfur content in jet fuel blends and making changes to engine design can also reduce exhaust particulates. Flight planning and changing the altitude to avoid condensate-producing environmental conditions are other possible strategies. However, routing changes can create traffic problems and contribute to CO_2 emissions by extending flights [3.122]. Powering aircraft by SAFs is an approach to reducing the impact of aviation on the climate by lowering aviation lifecycle CO_2 emissions and by reducing ice crystal numbers and radiative forcing from contrails [3.123]. Emissions affecting the climate and the differences between them are shown in **Figure 3.23** [3.94].

Figure 3.23 CO_2 and non-CO_2 emissions and the impact of SAF [3.94].

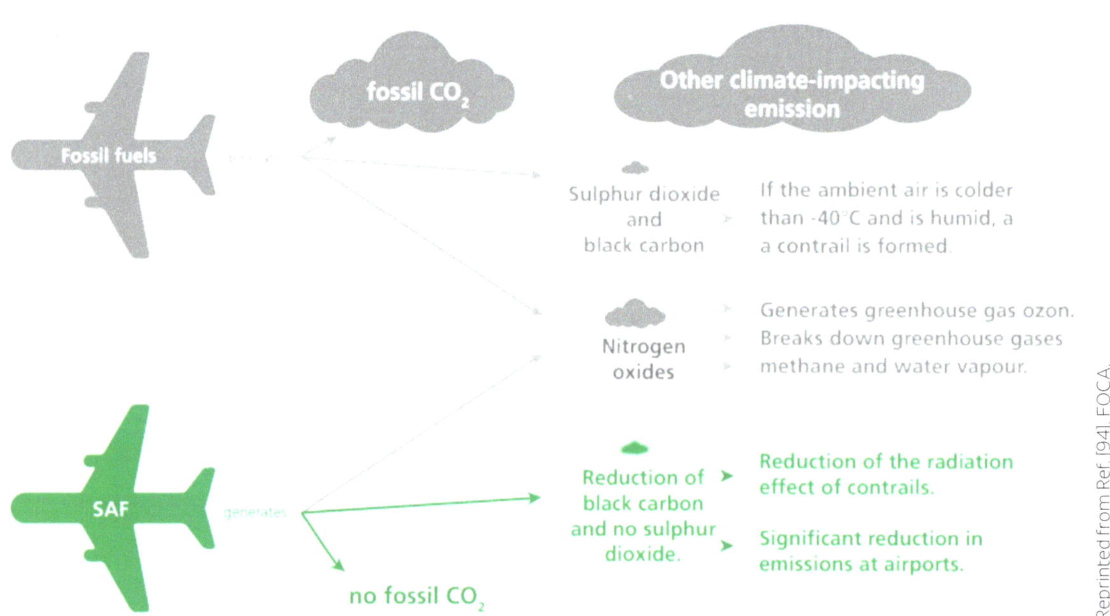

Currently, there is no sufficient literature support for the non-CO_2 emission characteristics of SAFs, which creates a significant research gap. Based on the studies examining the emissions by SAFs, the following emission reduction potentials can be listed.

CO_2 emissions of SAFs are significantly different from those of conventional aviation fuels from a lifecycle analysis perspective, and most SAFs can reduce CO_2 emissions by 41–89% [3.122, 3.123]. Compared to conventional aviation fuels, SAFs and blended fuels can significantly reduce SO_2 and PM emissions [3.124]. Pure FT-SPK can reduce SO_2 and PM emissions by 92% and 70–95%, respectively, due to its extremely low sulfur and aromatic compound content [3.125]. In contrast, the differences in NO_x emissions between the two fuel types are not significant since NO_x production mechanisms are largely due to thermal excitation and turbulent flow in the combustion chamber and emission performance is related to the power output and flame temperature profile in the engine test. There are minimal differences in NO_x emissions between 100% aviation kerosene, FT-SPK fuel, and HEFA fuel, which are not considered to be affected by changes in the SAF ratio in SAF/aviation kerosene blend fuels. The results show that there is minimal difference in NO_x emissions between different fuels, followed by CO_2 and CO, while the largest difference is observed in UHC emissions [3.124, 3.125, 3.126].

CO and UHC emissions are related to the engine operating conditions and physical/chemical properties of SAFs, and there is no significant upward or downward trend. If some properties of SAFs such as viscosity, density, and residence time do not match the atomization and combustion systems of the engine, incomplete combustion of the fuel occurs and, accordingly, an increase in UHC emissions occurs [3.127]. Studies found only fewer differences in CO emissions of SAFs compared to conventional aviation fuels; these differences are much less than the differences in SO_2 and PM emissions observed [3.126, 3.127, 3.128]. They contain fewer soot particles, thus significantly reducing contrail formation. Although the use of a pure (unblended) low-aromatic SAF is not currently permitted, studies have shown that SAFs can reduce the mass and number of emitted soot particles, potentially reducing the lifetime of condensate trail cirrus cloud [3.127, 3.129]. Even though the precise non-CO_2 effects are not yet assessed, these effects can be addressed using SAFs. In general, SAFs produce significantly lower SO_2 and PM emissions from aviation engine exhaust compared to conventional fuels. Emissions of SO_X also enhance contrail formation due to their coating effect on soot particles that are formed from the sulfur content in conventional jet fuel. Since neat SAF contains no sulfur, its use eliminates the effect of SO_X on contrail formation. According to previous studies, SAFs do not offer a significant advantage in CO, NO_x, and UHC emissions compared to conventional petroleum-based aviation fuels during engine combustion [3.125, 3.126]. CO, NO_x, and UHC emissions are determined by both the nature of the fuel and engine operating conditions, which explains the debate around the strength of the two fuels in terms of emissions of these pollutants [3.130]. From this perspective, combustion performance can be improved through fuel production and engine combustion technologies, and SAFs are promising fuel sources. In particular, SAFs produced from synthetic and biomass sources exhibit a significant advantage in CO_2 emissions from an LCA perspective, positioning them as an important strategy to address carbon emission challenges in the aviation industry [3.131]. In terms of encouraging research that will lead to the promotion of SAF use, electricity and hydrogen technologies, which are against SAFs, also have emission reduction projections. However, electric propulsion will eliminate all CO_2 and non-CO_2 emissions, while batteries will only be used for subregional aircraft. Such aircraft are capable of flying at altitudes where condensation trails do not form. Further advancements in battery technology are needed for the wider applicability of electric planes in aviation, limiting their contribution to decarbonization efforts by 2050. Hydrogen could theoretically service long-haul flights, but that would require an airframe redesign optimized for hydrogen use, which would not be feasible before 2050 [3.132]. This limits hydrogen's contribution to flights up to a maximum of 4000 km in range with retrofits to current aircraft designs. However, hydrogen comes with additional challenges, including sourcing (especially in the case of green hydrogen), transporting, and storing [3.133]. Thus, more research is needed due to storage and infrastructure problems compared to conventional jet fuel or SAFs.

Potential GHG Saving of SAFs

The aviation industry's reliance on petroleum-based fuels and the rising demand for air travel have led to a continuous increase in GHG emissions. Around 60% of CO_2 emissions from civil aviation worldwide are produced by aircraft with a gross take-off mass of over 45 tons and an average flight distance of over 6500 km [3.134]. BE, hydrogen, or ammonia propulsion engines, which are carbon-free alternatives to traditional jet engines, have limitations in terms of take-off weight and maximum range due to their lower energy content per unit mass [3.135]. Fully operational aircraft using these technologies are not expected to become viable alternatives for at least several decades. In addition, nuclear and/or renewable electricity generation capacities will need to be significantly increased to power turboelectric engines or produce green hydrogen so that CO_2 emissions from alternative technologies do not exceed those from fossil fuel kerosene operation [3.136]. The aviation sector, in particular, aims to achieve a 50% reduction in CO_2 emissions by 2050 compared to the 2005 levels.

To achieve this goal, the CO_2 reduction potential emerges by using SAFs, as shown in **Figure 3.24** [3.137]. SAFs are a solution to reduce fossil carbon emissions from aviation. CORSIA has been adopted globally to make and incentivize the production of SAFs economically competitive [3.138]. When produced from sustainably sourced biomass, it is assumed that emissions can be reduced throughout the fuel lifecycle. For the feedstocks given in **Table 3.3**, CO_2 saving provides a significant LCA in SAF production [3.138, 3.139, 3.140].

Figure 3.24 CO_2 emission reduction in international aviation [3.137].

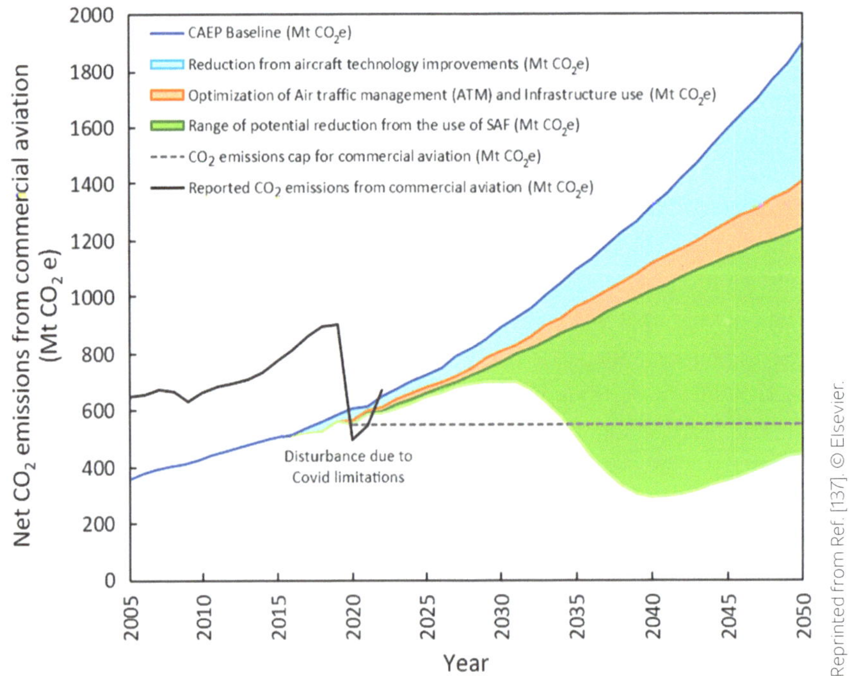

Table 3.3 CO_2 emission saving of SAFs [3.139, 3.140].

Fuel type	Pathway	Feedstock	gCO_2e/MJ	Emission saving (%)
Fossil fuel	Distillation	Petroleum	89	
SAFs	HEFA	UCO	13	85
		Jatropha	21	76
		Camelina	42	53
		Rapeseed	47	47
		Palm fatty acid distillate	20	77
	FT	Willow	-7	108
		Poplar	-6	107
		Corn stower	-3	103
		Forestry residues	−10	111
		MSW	5	94
		Agricultural residues	8	91
		Short-rotation woody crops	12	86
		Herbaceous energy crops	11	88
	ATJ	Corn stover	22	75
		Agricultural residues	29	67
		Forestry residues	24	73
		Switchgrass	30	66
		Herbaceous energy crops	44	51
	HTL	Forestry residues	20	78
	Pyrolysis	Forestry residues	30	67

The LCA evaluates the emission intensity of the investigated SAF production route and compares the net lifecycle emissions (including carbon capture technology) with the CORSIA benchmark values. To allow comparability with CORSIA, the functional unit is defined as gCO_2e/MJ. The raw materials and production methods shown in **Table 3.3** highlight the importance of CO_2 savings from SAF use. One way an airline can claim emission reductions is by using CORSIA-eligible fuels (CEFs). To be CORSIA-eligible, the fuel must be derived from sustainable feedstocks approved by the CAEP and produced using fuel conversion pathways accepted by ASTM [3.141]. These requirements make CEF and SAF similar concepts. The CORSIA program lists default lifecycle emission values for any CEFs. In the LCA perspective for SAFs, as shown in **Figure 3.25**, FT-SPK produced from forest residues is a promising fuel suitable for CORSIA and CEF [3.142]. The lifecycle fossil emissions reported were 49% higher than the CORSIA default value for FT-SPK. From an LCA perspective, SAFs have much lower CO_2 emissions than conventional aviation fuels.

Figure 3.25 LCA perspective for SAFs [3.142].

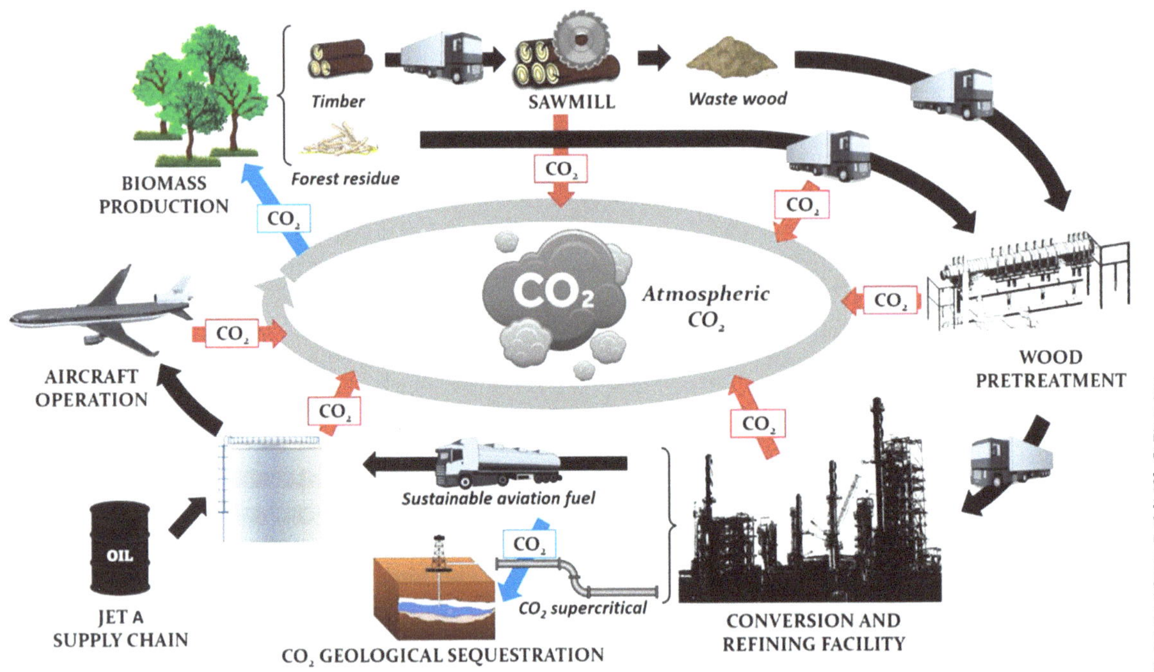

As shown in **Figure 3.26**, SAF blended at 1% and distributed equally across all transatlantic flights has been analyzed to reduce both the annual condensate footprint and total energy forcing (concentrates + change in CO_2 lifecycle emissions) by approximately 0.6% [3.143].

Accordingly, one way to reduce CO_2 emissions in aviation could perhaps be to make mandatory the minimum blend that SAFs will use as fuel so that passengers can be sure of the sustainability of the fuel their aircraft uses.

Figure 3.26 Using SAFs and climate benefit [3.143].

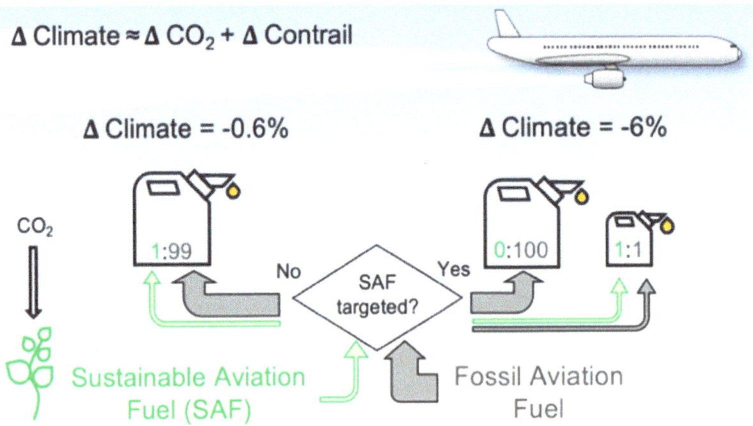

SAFs derived from different processes and feedstocks, such as FT, HEFA, AtJ, and pyrolysis, can significantly reduce GHG emissions (**Figure 3.27**) by 41–89% in the aviation industry [3.144].

SAFs are suitable for long-term storage due to their stability and low water solubility, making them well suited for use in fuel reserves. The process of blending with jet fuel is safe and creates a completely stable fuel, maintaining aviation's very high safety standards. A mix of low-carbon fuels and solutions, including SAFs, will be essential to decarbonize the aviation sector. The increased use of SAFs also allows the use of their waste products in a way that directly reduces GHG emissions by reducing air travel emissions. If this can be done on a large scale, it could rapidly transform the face of aviation from a highly polluting industry to an environmentally friendly one.

Figure 3.27 SAF CO_2 cycle [3.144].

References

3.1. Quante, G., Voß, S., Bullerdiek, N., Voigt, C. et al., "Hydroprocessing of Fossil Fuel-Based Aviation Kerosene – Technology Options and Climate Impact Mitigation Potentials," *Atmospheric Environment: X* 22 (2024): 100259, doi:https://doi.org/10.1016/j.aeaoa.2024.100259.

3.2. Baledón, M.S. and Kosoy, N., "'Problematizing' Carbon Emissions from International Aviation and the Role of Alternative Jet Fuels in Meeting ICAO's Mid-Century Aspirational Goals," *Journal of Air Transport Management* 71 (2018): 130-137, doi:https://doi.org/10.1016/j.jairtraman.2018.06.001.

3.3. Siddiqua, A., Hahladakis, J.N., and Al-Attiya, W.A.K.A., "An Overview of the Environmental Pollution and Health Effects Associated with Waste Landfilling and Open Dumping," *Environ Sci Pollut Res* 29 (2022): 58514-58536, doi:https://doi.org/10.1007/s11356-022-21578-z.

3.4. Dominic, D. and Baidurah, S., "A Review of Biological Processing Technologies for Palm Oil Mill Waste Treatment and Simultaneous Bioenergy Production at Laboratory Scale, Pilot Scale and Industrial Scale Applications with Technoeconomic Analysis," *Energy Conversion and Management: X* 26 (2025): 100914, doi:https://doi.org/10.1016/j.ecmx.2025.100914.

3.5. Güleç, F., Samson, A., Williams, O., Kostas, E.T. et al., "Biofuel Characteristics of Chars Produced from Rapeseed, Whitewood, and Seaweed via Thermal Conversion Technologies – Impacts of Feedstocks and Process Conditions," *Fuel Processing Technology* 238 (2022): 107492, doi:https://doi.org/10.1016/j.fuproc.2022.107492.

3.6. The Roundtable on Sustainable Biomaterials (RSB), "Sustainability Framework," accessed October 12, 2024, https://rsb.org/framework/#:~:text=RSB's%2012%20Principles%20%26%20Criteria%20(P%26C,by%20criteria%2C%20requirements%20and%20indicators.

3.7. The Roundtable on Sustainable Biomaterials (RSB), "Implementing Successful Public Policies to Shape the Sustainable Future of the Aviation Industry," accessed October 12, 2024, https://rsb.org/wp-content/uploads/2022/09/22-09-02-RSB-White-Paper_Print-D4-FINAL.pdf.

3.8. Thepchalerm, T. and Pinsuwan, S., "CEO Voices on Sustainable Aviation: An Analysis of Environmental Communication in the Airline Industry," *Green Technologies and Sustainability* 3, no. 3 (2025): 100194, doi:https://doi.org/10.1016/j.grets.2025.100194.

3.9. Acaroğlu, H., Güllü, M., Sivri, N., and Marquez, F.P.G., "How Can There Be an Economic Transition to a Green Ecosystem by Adapting Plastic-to-Fuel Technologies through Renewable Energy?" *Sustainable Energy Technologies and Assessments* 64 (2025): 103691, doi:https://doi.org/10.1016/j.seta.2024.103691.

3.10. Zhang, X., Shi, Y., Chen, Y., Hu, H. et al., "Ecosystem Simulation and Environmental Impact Analysis of Transforming Microalgae to Produce Jet Fuel," *Journal of Cleaner Production* 333 (2022): 130100, doi:https://doi.org/10.1016/j.jclepro.2021.130100.

3.11. Rogachuk, B.E. and Okolie, J.A., "Comparative Assessment of Pyrolysis and Gasification-Fischer Tropsch for Sustainable Aviation Fuel Production from Waste Tires," *Energy Conversion and Management* 302 (2024): 118110, doi:https://doi.org/10.1016/j.enconman.2024.118110.

3.12. Becken, S., Mackey, B., and Lee, D.S., "Implications of Preferential Access to Land and Clean Energy for Sustainable Aviation Fuels," *Science of the Total Environment* 886 (2023): 163883, doi:https://doi.org/10.1016/j.scitotenv.2023.163883.

3.13. PACE, "A Guide to ReFuel EU initiative for the Aviation Sector," accessed October 15, 2024, https://www.pace-esg.com/academy/a-guide-to-refuel-eu-initiative-for-the-aviation-sector/.

3.14. ICCT20, "Net-Zero Aviation: How It Started and How It's Going," accessed October 17, 2024, https://theicct.org/net-zero-aviation-how-it-started-and-how-its-going-apr24/.

3.15. Ember Energy, "World Passes 30% Renewable Electricity Milestone," accessed October 17, 2024, https://ember-energy.org/latest-updates/world-passes-30-renewable-electricity-milestone/.

3.16. The International Civil Aviation Organization, "Long-Term Aspiration Goal (LTAG)," accessed October 18, 2024, https://www.icao.int/environmental-protection/pages/SAF.aspx.

3.17. Jaén, R.L., Pupo, L.P., Lora, E.E.S., Eras, J.J.C. et al., "Assessing Biomass Production and Electricity Generation Potential in Current and Future Decarbonization Scenarios in Cuba Until 2050," *Energy Conversion and Management* 332 (2025): 119698, doi:https://doi.org/10.1016/j.enconman.2025.119698.

3.18. Wang, F. and Rijal, D., "Sustainable Aviation Fuels for Clean Skies: Exploring the Potential and Perspectives of Strained Hydrocarbons," *Energy Fuels* 38, no. 6 (2024): 4904-4920, doi:https://doi.org/10.1021/acs.energyfuels.3c04935.

3.19. Seymour, K., Held, M., Georges, G., and Boulouchos, K., "Fuel Estimation in Air Transportation: Modeling Global Fuel Consumption for Commercial Aviation," *Transportation Research Part D: Transport and Environment* 88 (2020): 102528, doi:https://doi.org/10.1016/j.trd.2020.102528.

3.20. Aerospace Technology Institute (ATI), "Non-CO_2 Technologies Roadmap," accessed October 18, 2024, https://www.ati.org.uk/wp-content/uploads/2024/03/ATI-Non-CO2-Technologies-Roadmap-Report-FINAL-March-2024.pdf.

3.21. Pagoni, I. and Psaraki-Kalouptsidi, V., "Calculation of Aircraft Fuel Consumption and CO_2 Emissions Based on Path Profile Estimation by Clustering and Registration," *Transportation Research Part D: Transport and Environment* 54 (2017): 172-190, doi:https://doi.org/10.1016/j.trd.2017.05.006.

3.22. Airbus, "Aviation's Road to Zero May Look Like This," accessed October 18, 2024, https://www.airbus.com/en/newsroom/news/2021-06-aviations-road-to-zero-may-look-like-this.

3.23. Jarin, J., Champion-Réaud, J., Sallinen, R., and Steenwinkel, E., "Emissions Comparison of 100% SAF With Bio-Aromatics and Conventional (Fossil) Jet Fuel," in *Proceedings of the ASME Turbo Expo 2024: Turbomachinery Technical Conference and Exposition. Volume 2: Ceramics and Ceramic Composites; Coal, Biomass, Hydrogen, and Alternative Fuels*, London, UK, June 24–28, 2024, https://doi.org/10.1115/GT2024-124002.

3.24. Kováčik, Ľ., Lusiak, T., and Novák, A., "Reducing Emissions from Aviation and Their Impact on Aviation Work in Agriculture," *Transportation Research Procedia* 55 (2021): 220-227, doi:https://doi.org/10.1016/j.trpro.2021.06.025.

3.25. Our World in Data, "Cars, Planes, Trains: Where Do CO_2 Emissions from Transport Come From?" accessed October 18, 2024, https://ourworldindata.org/co2-emissions-from-transport?&sznclid=51c9417b024946eb947fac4c7314e12f#article-citation.

3.26. Oldani, A.L., Solecki, A.E., and Lee, T., "Evaluation of Physicochemical Variability of Sustainable Aviation Fuels," *Front. Energy Res.* 10 (2022): 1052267, doi:https://doi.org/10.3389/fenrg.2022.1052267.

3.27. Skytender Solutions, "Climate-Neutral Aviation Takes Time, But Significant Improvements Are Already Possible," accessed October 18, 2024, https://skytendersolutions.com/climate-neutral-aviation-takes-time-but-significant-improvements-are-already-possible/.

3.28. Aerospace Technology Institute (ATI), "Aircraft Emissions and Their Climate Interactions," accessed October 18, 2024, https://www.ati.org.uk/wp-content/uploads/2024/03/ATI-Non-CO2-Technologies-Roadmap-Report-FINAL-March-2024.pdf.

3.29. Liang, Z., Liu, H., Fan, Y., Salehi, F. et al., "Applying Fischer Tropsch and Its Pentanol Blends Into an Aviation Compression Ignition Engine for PM Emissions Control," *Journal of the Energy Institute* 116 (2024): 101742, doi:https://doi.org/10.1016/j.joei.2024.101742.

3.30. Aerospace Technology Institute (ATI), "Non-CO_2 Emissions Comparison," accessed October 18, 2024, https://www.ati.org.uk/wp-content/uploads/2024/03/ATI-Non-CO2-Technologies-Roadmap-Report-FINAL-March-2024.pdf.

3.31. Roland Berger, "Aviation Emissions Reduction Roadmaps Assessed and Compared," accessed October 19, 2024, https://www.rolandberger.com/en/Insights/Publications/Aviation-emissions-reduction-roadmaps-assessed-and-compared.html.

3.32. Fay, C.D., Corcoran, B., and Diamond, D., "Green IoT Event Detection for Carbon-Emission Monitoring in Sensor Networks," *Sensors* 24, no. 1 (2024): 162, doi:https://doi.org/10.3390/s24010162.

3.33. Green Air, "European Aviation Coalition Urges Greater Support for SAF to Achieve EU Climate Targets," accessed October 19, 2024, https://www.greenairnews.com/?p=6727.

3.34. ECAC Guidance, "Sustainable Aviation Fuels (SAF)," accessed October 19, 2024, https://www.ecac-ceac.org/images/activities/environment/ECAC_Guidance_on_SAF_DGs_endorsed_HighRes.pdf.

3.35. NACO, "Sustainable Aviation and Climate Resilience," accessed October 21, 2024, https://www.naco.nl/en/expertise/sustainable-aviation-and-climate-resilience.

3.36. European Commission, "Fit for 55: Delivering on the Proposals," accessed October 21, 2024, https://commission.europa.eu/strategy-and-policy/priorities-2019-2024/european-green-deal/delivering-european-green-deal/fit-55-delivering-proposals_en.

3.37. NOW GmbH, "Renewable Energy Directive III (RED III)," accessed October 24, 2024, https://www.now-gmbh.de/wp-content/uploads/2024/01/Factsheet_REDIII.pdf.

3.38. GOV.UK, "CORSIA: How to Comply," accessed October 24, 2024, https://www.gov.uk/guidance/corsia-how-to-comply#:~:text=If%20you%20need%20to%20comply,April%20in%20the%20following%20year.

3.39. HFW, "Corsia Eligible Emission Units and an Airline's Offsetting Obligations," accessed October 25, 2024, https://www.hfw.com/insights/corsia-eligible-emission-units-and-an-airlines-offsetting-obligations/.

3.40. Wozny, F., Grimme, W., Maertens, S., and Scheelhaase, J., "CORSIA—A Feasible Second Best Solution?" *Applied Sciences* 12, no. 14 (2022): 7054, doi:https://doi.org/10.3390/app12147054.

3.41. NACO, "Sustainability throughout the Airport Ecosystem," accessed October 21, 2024, https://www.naco.nl/en/expertise/sustainable-aviation-and-climate-resilience.

3.42. GOV.UK, "Guidance Participating in the UK ETS," accessed October 22, 2024, https://www.gov.uk/government/publications/participating-in-the-uk-ets/participating-in-the-uk-ets.

3.43. Open Airlines, "Cleared for Compliance: Key Environmental Regulations for Airlines," accessed October 23, 2024, https://blog.openairlines.com/cleared-for-compliance-key-environmental-regulations-for-airlines.

3.44. Ruan, C., Chen, F., Cai, W., Qian, Y. et al., "Principles of Non-Intrusive Diagnostic Techniques and Their Applications for Fundamental Studies of Combustion Instabilities in Gas Turbine Combustors: A Brief Review," *Aerospace Science and Technology* 84 (2019): 585-603, doi:https://doi.org/10.1016/j.ast.2018.10.002.

3.45. Liu, Y., Sun, X., Sethi, V., Nalianda, D. et al., "Review of Modern Low Emissions Combustion Technologies for Aero Gas Turbine Engines," *Progress in Aerospace Sciences* 94 (2017): 12-45, doi:https://doi.org/10.1016/j.paerosci.2017.08.001.

3.46. The International Civil Aviation Organization (ICAO) Environmental Report, "Aircraft Technology Improvements," accessed October 24, 2024, https://www.icao.int/environmental-protection/Documents/EnvironmentReport-2010/ICAO_EnvReport10-Ch2_en.pdf.

3.47. Liu, Z., Hou, M., Sa, G., Wang, Y. et al., "Gas Turbine Multi-Working Conditions Identification and Performance Prediction Based on Deep Learning and Knowledge," *Energy* 308 (2024): 133011, doi:https://doi.org/10.1016/j.energy.2024.133011.

3.48. Xu, Z., Pei, J., Ding, S., Chen, L. et al., "Gas Exchange Optimization in Aircraft Engines Using Sustainable Aviation Fuel: A Design of Experiment and Genetic Algorithm Approach," *Energy and AI* 17 (2024): 100396, doi:https://doi.org/10.1016/j.egyai.2024.100396.

3.49. U.S. Department of Energy, "How Gas Turbine Power Plants Work," accessed October 25, 2024, https://www.energy.gov/fecm/how-gas-turbine-power-plants-work.

3.50. Ge Vernova, "What Is a Gas Turbine, and How Does It Work?" accessed October 25, 2024, https://www.gevernova.com/gas-power/resources/education/what-is-a-gas-turbine.

3.51. Inspenet, "Principles and Characteristics of Gas Turbines," accessed October 25, 2024, https://inspenet.com/en/articulo/gas-turbine-characteristics-principles/.

3.52. Altintas, F., "Irreversible Quantum Carnot Cycle and Its Comparison with Quantum Otto Cycle," *Physics Letters A* 533 (2025): 130212, doi:https://doi.org/10.1016/j.physleta.2024.130212.

3.53. Kawasaki, "Principle of Gas Turbine Operation," accessed October 26, 2024, https://global.kawasaki.com/en/energy/equipment/gas_turbines/outline.html.

3.54. Plot Who Ask Why, "How Does a Jet Engine Work? Understanding the 6 Main Types," accessed October 26, 2024, https://pilotswhoaskwhy.com/2024/04/21/how-does-a-jet-engine-work-understanding-the-6-main-types/.

3.55. Aerospace Notes, "Gas Turbine," accessed October 26, 2024, https://aerospacenotes.com/propulsion-1/gas-turbine/.

3.56. Turbines Info, "Gas Turbines," accessed October 26, 2024, https://www.turbinesinfo.com/gas-turbines/.

3.57. Zohuri, B., "Gas Turbine Working Principles," in: *Combined Cycle Driven Efficiency for Next Generation Nuclear Power Plants* (Cham, Switzerland: Springer, 2015), https://doi.org/10.1007/978-3-319-15560-9_7.

3.58. Mohamed, O. and Khalil, A., "Progress in Modeling and Control of Gas Turbine Power Generation Systems: A Survey," *Energies* 13, no. 9 (2020): 2358, doi:https://doi.org/10.3390/en13092358.

3.59. Roumeliotis, I., Aretakis, N., and Alexiou, A., "Industrial Gas Turbine Health and Performance Assessment with Field Data," *J. Eng. Gas Turbines Power* 139, no. 5 (2017): 051202, doi:https://doi.org/10.1115/1.4034986.

3.60. Wei, S., Zhang, Z., Wu, L., Sun, L. et al., "Combustion Characteristics of RP-3 Aviation Kerosene/n-Butanol Blended Fuel in a Compression Ignition Engine," *Journal of the Energy Institute* 115 (2024): 101675, doi:https://doi.org/10.1016/j.joei.2024.101675.

3.61. Manigandan, S., Atabani, A.E., Ponnusamy, V.K., and Gunasekar, P., "Impact of Additives in Jet-A Fuel Blends on Combustion, Emission and Exergetic Analysis Using a Micro-Gas Turbine Engine," *Fuel* 276 (2020): 118104, doi:https://doi.org/10.1016/j.fuel.2020.118104.

3.62. Raji, A.M., Manescau, B., Chetehouna, K., Ango, S.E. et al., "Performance and Spray Characteristics of Fossil JET A-1 and bioJET Fuel: A Comprehensive Review," *Renewable and Sustainable Energy Reviews* 207 (2025): 114970, doi:https://doi.org/10.1016/j.rser.2024.114970.

3.63. Shafiee, P., Dorneanu, B., and Arellano-Garcia, H., "Improving Catalysts and Operating Conditions Using Machine Learning in Fischer-Tropsch Synthesis of Jet Fuels (C8-C16)," *Chemical Engineering Journal Advances* 21 (2025): 100702, doi:https://doi.org/10.1016/j.ceja.2024.100702.

3.64. Cowart, J., Foley, M.P., and Prak, D.L., "The Development and Testing of Navy Jet Fuel (JP-5) Surrogates," *Fuel* 249 (2019): 80-88, doi:https://doi.org/10.1016/j.fuel.2019.03.096.

3.65. Liang, Y., Liu, X., Yang, M., Hui, X. et al., "Investigating the Oxidation Characteristic of a Hydro-Processed Bio-Jet Fuel: Experimental and Modeling Study," *Combustion and Flame* 270 (2024): 113778, doi:https://doi.org/10.1016/j.combustflame.2024.113778.

3.66. Paulsen, M.M., Petersen, S.B., Lozano, E.M., and Pedersen, T.H., "Techno-Economic Study of Integrated High-Temperature Direct Air Capture with Hydrogen-Based Calcination and Fischer–Tropsch Synthesis for Jet Fuel Production," *Applied Energy* 369 (2024): 123524, doi:https://doi.org/10.1016/j.apenergy.2024.123524.

3.67. The International Civil Aviation Organization (ICAO) Environment, "Update on 100% SAF Testing and Recent Achievements," accessed November 7, 2024, https://www.icao.int/environmental-protection/Documents/ACT-SAF/ACT-SAF%20Series%2014%20-%20Updates%20on%20100%20per%20cent%20SAF.pdf.

3.68. BP, "Sustainable Aviation Fuel (SAF) – Specifications and Composition," accessed November 7, 2024, https://www.bp.com/en/global/air-bp/news-and-views/views/saf_specs_and_comparison.html.

3.69. Ebrahimi, A., Rolt, A., Jafari, S., and Anton, J.H., "A Review on Liquid Hydrogen Fuel Systems in Aircraft Applications for Gas Turbine Engines," *International Journal of Hydrogen Energy* 91 (2024): 88-105, doi:https://doi.org/10.1016/j.ijhydene.2024.10.121.

3.70. Khandelwal, B., Karakurt, A., Sekaran, P.R., Sethi, V. et al., "Hydrogen Powered Aircraft : The Future of Air Transport," *Progress in Aerospace Sciences* 60 (2013): 45-59, doi:https://doi.org/10.1016/j.paerosci.2012.12.002.

3.71. Teixeira, A.T., da Silva, A.C.M., Cavalcante, R.M., and Young, A.F., "Process Simulation and Economic Evaluation of the Alcohol-to-Jet Production of Sustainable Aviation Fuel in the Brazilian Context," *Energy Conversion and Management* 319 (2024): 118947, doi:https://doi.org/10.1016/j.enconman.2024.118947.

3.72. Benrabah, R., El Sayah, Z., Le, M.D., Warren, Y.A.D. et al., "Experimental Study of the Impact of Alcohols on the Oxidation Stability of a Surrogate Jet-Fuel," *Fuel* 361 (2024): 130750, doi:https://doi.org/10.1016/j.fuel.2023.130750.

3.73. Pandey, V., Badruddin, I.A., Terfasa, T.T., Tesfamariam, B.B. et al., "Experimental Investigation of the Impact of CeO_2 Nanoparticles in Jet-A and Jatropha-SPK Blended Fuel in an Aircraft Can-Combustor at Flight Conditions," *Fuel* 317 (2022): 123393, doi:https://doi.org/10.1016/j.fuel.2022.123393.

3.74. Rye, L. and Wilson, C., "The Influence of Alternative Fuel Composition on Gas Turbine Ignition Performance," *Fuel* 96 (2012): 277-283, doi:https://doi.org/10.1016/j.fuel.2011.12.047.

3.75. Yang, L., Yang, Y., Zhou, Y., and Shi, X., "Research the Synergistic Carbon Reduction Effects of Sulfur Dioxide Emissions Trading Policy," *Journal of Cleaner Production* 447 (2024): 141483, doi:https://doi.org/10.1016/j.jclepro.2024.141483.

3.76. Won, S.H., Veloo, P.S., Dooley, S., Santner, J. et al., "Predicting the Global Combustion Behaviors of Petroleum-Derived and Alternative Jet Fuels by Simple Fuel Property Measurements," *Fuel* 168 (2016): 34-46, doi:https://doi.org/10.1016/j.fuel.2015.11.026.

3.77. Aweid, R.S., Mustafa, A.N., Ali, O.M., and Ali, B.M., "Thermal, Environmental, and Economic Analysis of the Gas Turbine Fogging System," *Results in Engineering* 25 (2025): 103952, doi:https://doi.org/10.1016/j.rineng.2025.103952.

3.78. Lown, A.L., Peereboom, L., Mueller, S.A., Anderson, J.E. et al., "Cold Flow Properties for Blends of Biofuels with Diesel and Jet Fuels," *Fuel* 117 (2014): 544-551, doi:https://doi.org/10.1016/j.fuel.2013.09.067.

3.79. Chen, X., Wang, H., Wang, X., Wang, N. et al., "Experimental and Numerical Investigations into Cold Flow Characteristics of Multiple Micro-Mixing Jets for Hydrogen-Rich Gas Turbines," *Aerospace Science and Technology* 155 (2024): 109702, doi:https://doi.org/10.1016/j.ast.2024.109702.

3.80. Gardi, A., Sabatini, R., and Ramasamy, S., "Multi-Objective Optimisation of Aircraft Flight Trajectories in the ATM and Avionics Context," *Progress in Aerospace Sciences* 83 (2016): 1-36, doi:https://doi.org/10.1016/j.paerosci.2015.11.006.

3.81. Cui, Q., Lei, Y.-l., Jia, Z.-k., Wang, Y. et al., "Path Analysis for Controlling Climate Change in Global Aviation," *iScience* 27, no. 6 (2024): 110126, doi:https://doi.org/10.1016/j.isci.2024.110126.

3.82. NASA, "Aircraft Gas Turbine Combustor Emissions Research Past, Present, and Future," accessed November 8, 2024, https://ntrs.nasa.gov/api/citations/20230011949/downloads/Hicks_UCIrvine2023SummerSchoolLecture_YRHFinal.pdf.

3.83. Zou, R., Wang, B., Wang, K., Shang, W.-L. et al., "A Pathway to Sustainable Aviation: Modeling Aircraft Takeoff Mass for Precise Fuel Consumption and Aircraft Emission Calculations," *Energy* 319 (2025): 135074, doi:https://doi.org/10.1016/j.energy.2025.135074.

3.84. Financial Times, "Airlines Face Long-Haul to Carbon-Free Flying," accessed November 8, 2024, https://www.ft.com/content/1081c524-0c73-11ea-bb52-34c8d9dc6d84.

3.85. CE Delft, "Potential for Reducing Aviation Non-CO_2 Emissions through Cleaner Jet Fuel," accessed November 15, 2024, https://cedelft.eu/wp-content/uploads/sites/2/2022/03/CE_Delft_210410_Potential_reducing_aviation_non-CO2_emissions_cleaner_jet_fuel_FINAL.pdf.

3.86. Loo, B.P.Y., Li, L., Psaraki, V., and Pagoni, I., "CO_2 Emissions Associated with Hubbing Activities in Air Transport: An International Comparison," *Journal of Transport Geography* 34 (2014): 185-193, doi:https://doi.org/10.1016/j.jtrangeo.2013.12.006.

3.87. Chao, C.-C., "Assessment of Carbon Emission Costs for Air Cargo Transportation," *Transportation Research Part D: Transport and Environment* 33 (2014): 186-195, doi:https://doi.org/10.1016/j.trd.2014.06.004.

3.88. RMI, "Contrail Formation and Warming Effects," accessed November 15, 2024, https://rmi.org/aviation-contrails-what-we-know-and-what-we-dont-about-this-warming-phenomenon/.

3.89. American Airlines, "Contrail Avoidance," accessed November 15, 2024, https://news.aa.com/esg/climate-change/contrail-avoidance/.

3.90. Quante, G., Enderle, B., Laybourn, P., Holm, P.W. et al., "Segregated Supply of Sustainable Aviation Fuel to Reduce Contrail Energy Forcing – Demonstration and Potentials," *Journal of the Air Transport Research Society* 4 (2025): 100049, doi:https://doi.org/10.1016/j.jatrs.2024.100049.

3.91. GE Power Systems, "Gas Turbine Emissions and Control," accessed November 15, 2024, https://www.gevernova.com/content/dam/gepower-new/global/en_US/downloads/gas-new-site/resources/reference/ger-4211-gas-turbine-emissions-and-control.pdf.

3.92. Cui, Q. and Zhou, Y., "Protocol to Calculate the Climate Effects of the Global Aviation Industry Using the Aviation-FAIR Model," *STAR Protocols* 6, no. 1 (2025): 103656, doi:https://doi.org/10.1016/j.xpro.2025.103656.

3.93. Scheelhaase, J., Grimme, W., and Maertens, S., "EU Trilogue Results for the Aviation Sector – Key Issues and Expected Impacts," *Transportation Research Procedia* 78 (2024): 206-214, doi:https://doi.org/10.1016/j.trpro.2024.02.027.

3.94. Switzerland Federal Office of Civil Aviation FOCA, "CO_2 and Non-CO_2 Emissions Produced by Aviation," accessed November 19, 2024, https://www.bazl.admin.ch/bazl/en/home/themen/umwelt/klima/emissionen/co2_nicht-co2_emissionen.html.

3.95. Yu-Jie, H., Yang, L., Cui, H., Wang, H. et al., "Developing a Balanced Strategy: A Multi-Objective Model for Emissions Reduction and Development of Civil Aviation in China," *Energy* 307 (2024): 132597, doi:https://doi.org/10.1016/j.energy.2024.132597.

3.96. Gunerhan, A., Altuntas, O., and Caliskan, H., "Analyzing the Influence of Feedstock Selection in Pyrolysis on Aviation Gas Turbine Engines: A Study on Performance, Combustion Efficiency, and Emission Profiles," *Energy* 306 (2024): 132513, doi:https://doi.org/10.1016/j.energy.2024.132513.

3.97. Wang, Y. and Sohn, C.H., "Effects of Swirl Premixed Flame Interaction on NOx Emission in a Gas Turbine Combustor with Multi-Burners," *Journal of the Energy Institute* 109 (2023): 101274, doi:https://doi.org/10.1016/j.joei.2023.101274.

3.98. Okafor, E.C., Kurata, O., Yamashita, H., Iki, N. et al., "Achieving High Flame Stability with Low NO and Zero N_2O and NH_3 Emissions during Liquid Ammonia Spray Combustion with Gas Turbine Combustors," *Proceedings of the Combustion Institute* 40, no. 1–4 (2024): 105340, doi:https://doi.org/10.1016/j.proci.2024.105340.

3.99. Bendtsen, K.M., Bengtsen, E., Saber, A.T. et al., "A Review of Health Effects Associated with Exposure to Jet Engine Emissions in and around Airports," *Environ Health* 20 (2021): 10, doi:https://doi.org/10.1186/s12940-020-00690-y.

3.100. Abdalla, M.S.M., Balli, O., Adali, O.H., Korba, P. et al., "Thermodynamic, Sustainability, Environmental and Damage Cost Analyses of Jet Fuel Starter Gas Turbine Engine," *Energy* 267 (2023): 126487, doi:https://doi.org/10.1016/j.energy.2022.126487.

3.101. Zhang, C., Chen, L., Ding, S., Zhou, X. et al., "Mitigation Effects of Alternative Aviation Fuels on Non-Volatile Particulate Matter Emissions from Aircraft Gas Turbine Engines: A Review," *Science of the Total Environment* 820 (2022): 153233, doi:https://doi.org/10.1016/j.scitotenv.2022.153233.

3.102. Melzi, G., van Triel, J., Durand, E., Crayford, A. et al., "Toxicological Evaluation of Primary Particulate Matter Emitted from Combustion of Aviation Fuel," *Chemosphere* 363 (2024): 142958, doi:https://doi.org/10.1016/j.chemosphere.2024.142958.

3.103. Oberthür, S., "Institutional Interaction to Address Greenhouse Gas Emissions from International Transport: ICAO, IMO and the Kyoto Protocol," *Climate Policy* 3, no. 3 (2023): 191-205, doi:https://doi.org/10.1016/S1469-3062(03)00060-3.

3.104. Ekici, S., Ayar, M., and Karakoc, T.H., "Fuel-Saving and Emission Accounting: An Airliner Case Study for Green Engine Selection," *Energy* 282 (2023): 128922, doi:https://doi.org/10.1016/j.energy.2023.128922.

3.105. Adolfo, D., Bertini, D., Gamannossi, A., and Carcasci, C., "Thermodynamic Analysis of an Aircraft Engine to Estimate Performance and Emissions at LTO Cycle," *Energy Procedia* 126 (2017): 915-922, doi:https://doi.org/10.1016/j.egypro.2017.08.162.

3.106. Inoue, C., Higaki, S., Fujii, H., Ishikawa, N. et al., "On-Site Direct Imaging of Carbon Dioxide Emissions from Aircraft during Landing and Take-Off Cycle," *Environmental Advances* 13 (2023): 100432, doi:https://doi.org/10.1016/j.envadv.2023.100432.

3.107. Pan, H., Zhou, X., Xie, S., Zhicai, D. et al., "Selective Production of Monocyclic Aromatic Hydrocarbon from Agricultural Waste Wheat Straw for Aviation Fuel Using Ni/ZSM-5 Catalyst," *Biomass and Bioenergy* 165 (2022): 106592, doi:https://doi.org/10.1016/j.biombioe.2022.106592.

3.108. Delbecq, S., Fontane, J., Gourdain, N., Planès, T. et al., "Sustainable Aviation in the Context of the Paris Agreement: A Review of Prospective Scenarios and Their Technological Mitigation Levers," *Progress in Aerospace Sciences* 141 (2023): 100920, doi:https://doi.org/10.1016/j.paerosci.2023.100920.

3.109. Turgut, E.T., Usanmaz, O., and Rosen, M.A., "Empirical Analysis of the Effect of Descent Flight Path Angle on Primary Gaseous Emissions of Commercial Aircraft," *Environmental Pollution* 236 (2018): 226-235, doi:https://doi.org/10.1016/j.envpol.2018.01.084.

3.110. Zhang, F., Luo, L., Zhang, J., Cai, W. et al., "A Strategy for Enhancing Ignition of Aviation Fuels at High Altitudes Using Nanoparticle and Fuel-Soluble Catalysts," *Aerospace Science and Technology* 161 (2025): 110160, doi:https://doi.org/10.1016/j.ast.2025.110160.

3.111. Pelucchi, M., Oßwald, P., Pejpichestakul, W., Frassoldati, A. et al., "On the Combustion and Sooting Behavior of Standard and Hydro-Treated Jet Fuels: An Experimental and Modeling Study on the Compositional Effects," *Proceedings of the Combustion Institute* 38, no. 1 (2021): 523-532, doi:https://doi.org/10.1016/j.proci.2020.06.353.

3.112. Yan, B., Lu, S., Zhou, S., Chen, G. et al., "Insights into the Role of Fuel Structure and Blending in Soot Formation for C1–C4 Hydrocarbon Flames," *Journal of the Energy Institute* 119 (2025): 102011, doi:https://doi.org/10.1016/j.joei.2025.102011.

3.113. Diao, X., Xiong, Y., Shi, Y., Ma, L. et al., "Catalytic Hydrodeoxygenation and C–C Coupling of Lignin and Its Derivatives into Renewable Jet-Fuel-Range Cycloalkanes," *Green Chemistry* 26, no. 23 (2024): 11406-11426, doi:https://doi.org/10.1039/d4gc02051k.

3.114. Eun, J., Insyani, R., Choi, J.-W., Suh, D.J. et al., "Production of High-Carbon-Number Naphthenes for Bio-Aviation Fuels from Bio-Crude Prepared by Fast Pyrolysis of Lignocellulose," *Energy Conversion and Management* 314 (2024): 118696, doi:https://doi.org/10.1016/j.enconman.2024.118696.

3.115. Hart, A., Onwudili, J.A., Yildirir, E., and Hashemnezhad, S.E., "Energy-Dense Sustainable Aviation Fuel-Range Hydrocarbons from Cyclohexanone as a Biomass-Derived Feedstock via Sequential Catalytic Aldol Condensation and Hydrodeoxygenation," *Chemical Engineering Journal* 509 (2025): 161494, doi:https://doi.org/10.1016/j.cej.2025.161494.

3.116. Bernabei, M., Reda, R., Galiero, R., and Bocchinfuso, G., "Determination of Total and Polycyclic Aromatic Hydrocarbons in Aviation Jet Fuel," *Journal of Chromatography A* 985, no. 1–2 (2003): 197-203, doi:https://doi.org/10.1016/S0021-9673(02)01826-5.

3.117. Chang, Y., Huang, S., Li, H., and Jia, M., "Investigation of the Criteria for Construction of Reliable Surrogate Fuels: A Case Study of Jet-A Aviation Kerosene," *Combustion and Flame* 272 (2025): 113866, doi:https://doi.org/10.1016/j.combustflame.2024.113866.

3.118. Khan, M.Z.A., Khan, H.A., Ravi, S.S., Turner, J.W.G. et al., "Potential of Clean Liquid Fuels in Decarbonizing Transportation – An Overlooked Net- Zero Pathway?" *Renewable and Sustainable Energy Reviews* 183 (2023): 113483, doi:https://doi.org/10.1016/j.rser.2023.113483.

3.119. The International Civil Aviation Organization (ICAO), "Committee on Aviation Environmental Protection (CAEP)," accessed December 2, 2024, https://www.icao.int/environmental-protection/pages/caep.aspx.

3.120. Medlin, R., Meeks, S., Vasel-Be-Hagh, A., Damazo, J. et al., "Ammonia versus Kerosene Contrails: A Review," *Progress in Aerospace Sciences* 153 (2025): 101074, doi:https://doi.org/10.1016/j.paerosci.2024.101074.

3.121. Klenner, J., Lund, M.T., Muri, H., and Strømman, A.H., "Emission Location Affects Impacts on Atmosphere and Climate from Alternative Fuels for Norwegian Domestic Aviation," *Atmospheric Environment: X* 24 (2024): 100301, doi:https://doi.org/10.1016/j.aeaoa.2024.100301.

3.122. Zhao, Y., Zhu, M., Zhou, L., Gao, M. et al., "Evaluating High-Resolution Aviation Emissions Using Real-Time Flight Data," *Science of the Total Environment* 951 (2024): 175429, doi:https://doi.org/10.1016/j.scitotenv.2024.175429.

3.123. Gourgue, N., Boucher, O., and Barthès, L., "A Dataset of Annotated Ground-Based Images for the Development of Contrail Detection Algorithms," *Data in Brief* 59 (2025): 111364, doi:https://doi.org/10.1016/j.dib.2025.111364.

3.124. Lee, D.S., Allen, M.R., Cumpsty, N., Owen, B. et al., "Uncertainties in Mitigating Aviation Non-CO_2 Emissions for Climate and Air Quality Using Hydrocarbon Fuels," *Environmental Science: Atmospheres* 3, no. 12 (2023): 1693-1740, doi:https://doi.org/10.1039/d3ea00091e.

3.125. Ge, F., Yu, Z., Li, Y., Zhu, M. et al., "Predicting Aviation Non-Volatile Particulate Matter Emissions at Cruise via Convolutional Neural Network," *Science of the Total Environment* 850 (2022): 158089, doi:https://doi.org/10.1016/j.scitotenv.2022.158089.

3.126. Chen, L., Gao, M., Zhu, M., Zhong, S. et al., "Towards Precision Aviation Emission Modeling: A Hybrid Paradigm of Convolutional Neural Networks and Semi-Empirical Formulas for Full Flight Phase Gas Pollutant Indices," *Science of the Total Environment* 957 (2024): 177414, doi:https://doi.org/10.1016/j.scitotenv.2024.177414.

3.127. Scheelhaase, J., Maertens, S., and Grimme, W., "Options for Improving the EU Emissions Trading Scheme (EU ETS) for Aviation," *Transportation Research Procedia* 59 (2021): 193-202, doi:https://doi.org/10.1016/j.trpro.2021.11.111.

3.128. Gaillot, T., Beauchet, S., Lorne, D., and Krim, L., "The Impact of Fossil Jet Fuel Emissions at Altitude on Climate Change: A Life Cycle Assessment Study of a Long-Haul Flight at Different Time Horizons," *Atmospheric Environment* 311 (2023): 119983, doi:https://doi.org/10.1016/j.atmosenv.2023.119983.

3.129. Lim, Y., Gardi, A., and Sabatini, R., "Optimal Aircraft Trajectories to Minimize the Radiative Impact of Contrails and CO_2," *Energy Procedia* 110 (2017): 446-452, doi:https://doi.org/10.1016/j.egypro.2017.03.167.

3.130. Ahlström, J., Jafri, Y., Wetterlund, E., and Furusjö, E., "Sustainable Aviation Fuels – Options for Negative Emissions and High Carbon Efficiency," *International Journal of Greenhouse Gas Control* 125 (2023): 103886, doi:https://doi.org/10.1016/j.ijggc.2023.103886.

3.131. Bell, A., Mannion, L.A., Kelly, M., Ghaani, M.R. et al., "Life Cycle CO_2e Intensity of Commercial Aviation with Specific Sustainable Aviation Fuels," *Applied Energy* 382 (2025): 125075, doi:https://doi.org/10.1016/j.apenergy.2024.125075.

3.132. Oesingmann, K., Grimme, W., and Scheelhaase, J., "Hydrogen in Aviation: A Simulation of Demand, Price Dynamics, and CO_2 Emission Reduction Potentials," *International Journal of Hydrogen Energy* 64 (2024): 633-642, doi:https://doi.org/10.1016/j.ijhydene.2024.03.241.

3.133. Manigandan, S., Praveenkumar, T.R., Ryu, J.I., Verma, T.N. et al., "Role of Hydrogen on Aviation Sector: A Review on Hydrogen Storage, Fuel Flexibility, Flame Stability, and Emissions Reduction on Gas Turbines Engines," *Fuel* 352 (2023): 129064, doi:https://doi.org/10.1016/j.fuel.2023.129064.

3.134. Wang, X., Guo, L., Lv, J., Li, M. et al., "Process Design, Modeling and Life Cycle Analysis of Energy Consumption and GHG Emission for Jet Fuel Production from Bioethanol in China," *Journal of Cleaner Production* 389 (2023): 136027, doi:https://doi.org/10.1016/j.jclepro.2023.136027.

3.135. Baumeister, S., Simić, T.K., and Ganić, E., "Emissions Reduction Potentials in Business Aviation with Electric Aircraft," *Transportation Research Part D: Transport and Environment* 136 (2024): 104415, doi:https://doi.org/10.1016/j.trd.2024.104415.

3.136. Seber, G., Escobar, N., Valin, H., and Malina, R., "Uncertainty in Life Cycle Greenhouse Gas Emissions of Sustainable Aviation Fuels from Vegetable Oils," *Renewable and Sustainable Energy Reviews* 170 (2022): 112945, doi:https://doi.org/10.1016/j.rser.2022.112945.

3.137. Zhang, C., Fu, R., Kang, L., Ma, Y. et al., "An Upcycling Bioprocess for Sustainable Aviation Fuel Production from Food Waste-Derived Greenhouse Gases: Life Cycle Assessment and Techno-Economic Analysis," *Chemical Engineering Journal* 486 (2024): 150242, doi:https://doi.org/10.1016/j.cej.2024.150242.

3.138. Yan, W., Cheng, F., Wang, X., and Zhang, C., "Estimating the Potential of Sustainable Aviation Fuel Enabled Aviation Decarbonization in China: An Integrated Resource-Technology-Economy-Environment Analysis Framework," *Computer Aided Chemical Engineering* 53 (2024): 2371-2376, doi:https://doi.org/10.1016/B978-0-443-28824-1.50396-3.

3.139. Staples, M.D., Malina, R., Suresh, P., Hileman, J.I. et al., "Aviation CO_2 Emissions Reductions from the Use of Alternative Jet Fuels," *Energy Policy* 114 (2018): 342-354, doi:https://doi.org/10.1016/j.enpol.2017.12.007.

3.140. Partnership for Air Transportation, "Life Cycle Greenhouse Gas Emissions from Alternative Jet Fuels," accessed December 28, 2024, https://web.mit.edu/aeroastro/partner/reports/proj28/partner-proj28-2010-001.pdf.

3.141. Ahmad, S., Ouenniche, J., Kolosz, B.W., Greening, P. et al., "A Stakeholders' Participatory Approach to Multi-Criteria Assessment of Sustainable Aviation Fuels Production Pathways," *International Journal of Production Economics* 238 (2021): 108156, doi:https://doi.org/10.1016/j.ijpe.2021.108156.

3.142. Almena, A., Siu, R., Chong, K., Thornley, P. et al., "Reducing the Environmental Impact of International Aviation through Sustainable Aviation Fuel with Integrated Carbon Capture and Storage," *Energy Conversion and Management* 303 (2024): 118186, doi:https://doi.org/10.1016/j.enconman.2024.118186.

3.143. Teoh, R., Schumann, U., Voigt, C., Schripp, T. et al., "Targeted Use of Sustainable Aviation Fuel to Maximize Climate Benefits," *Environmental Science & Technology* 56, no. 23 (2022): 17246-17255, doi:https://doi.org/10.1021/acs.est.2c05781.

3.144. Malaysian Investment Development Authority (MIDA), "Greener Flights for the Greater Good," accessed December 28, 2024, https://www.mida.gov.my/greener-flights-for-the-greater-good/.

Chapter 04

Usage of SAFs

Decarbonizing the aviation sector requires a multipronged approach in terms of alternative fuel use and efficiency, to meet emission reductions and sustainability requirements and goals [4.1]. Inconsistent regulations worldwide have been a challenge for most airlines operating globally. Therefore, additional guidelines and commitments from global organizations such as IATA, ICAO, and the Air Transport Action Group (ATAG) also compel their members to reduce CO_2 emissions [4.2]. In order for ICAO and IATA to achieve their net-zero emissions target by 2050, a greater global emphasis is being placed on the development of SAF technology and its practical application in existing aircraft technology [4.3]. Aircraft manufacturers can increase fuel economy by approximately 20% with the advanced combustion and flight systems implemented in every new generation of aircraft [4.4]. However, achieving lower fuel consumption is not sufficient to meet the environmental and sustainability goals of international agreements, as it will not change the formation of end-of-combustion pollutants due to fuel properties. Thus, the use of SAF is being promoted by many airlines and national organizations that have made various commitments to decarbonization [4.5].

Participation in the CORSIA Pilot Phase and Phase 1 is voluntary for countries, which regulate emissions from flights to and from airports [4.6]. From Phase 2 onward, participation becomes mandatory for most ICAO member countries, with a few exceptions. Many countries have voluntarily undertaken early implementation ahead of the internationally established target requirement to support the use of SAFs. As shown in **Figure 4.1**, as of November 2024,

129 countries have voluntarily agreed to participate in Phase 1 [4.7]. To achieve the goals of the Paris Climate Change Agreement, all sectors of the global economy need to play a role in reducing GHG emissions. Although countries with lower economic status and limited air traffic are exempt, many of them are among those that have chosen to participate in the policies adopted for the use of SAFs.

Figure 4.1 CORSIA implementation by country [4.7].

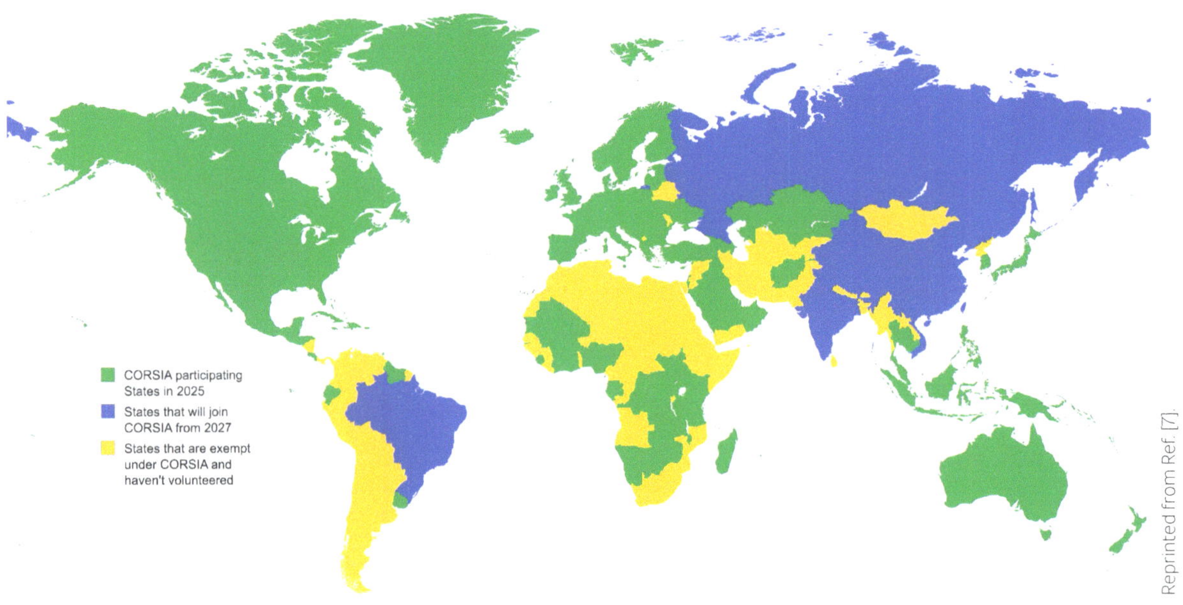

Reprinted from Ref. [7].

The use of SAFs has become widespread since the first test flight on a commercial aircraft in 2008 and the approval of SAFs for use in aircraft operations in 2011. SAFs are now produced every day and used in commercial flights. The compliance process for CORSIA began in 2019 for airlines with annual emissions exceeding 10,000 tons. Safe flights between voluntary countries began in 2021 due to offset requirements met through various activities, with full implementation expected in 2027 [4.8, 4.9]. IATA, a trade association representing 330 airlines worldwide, launched the Fly Net Zero initiative in 2021 to achieve net-zero carbon by 2050 [4.10]. The use of SAFs is assessed to represent 65% in achieving the CO_2 emission reduction target by 2050, with the rest of the reductions coming from the following: offsets/carbon capture (19%), development of new technologies (13%), and the improvement of infrastructure/operations (3%). Multiple levers can be used in different combinations to achieve net-zero emissions. **Figure 4.2** shows worldwide statistics regarding the use of SAFs, which are the most effective solutions according to these combinations [4.11].

Figure 4.2 SAF usage statistics [4.11].

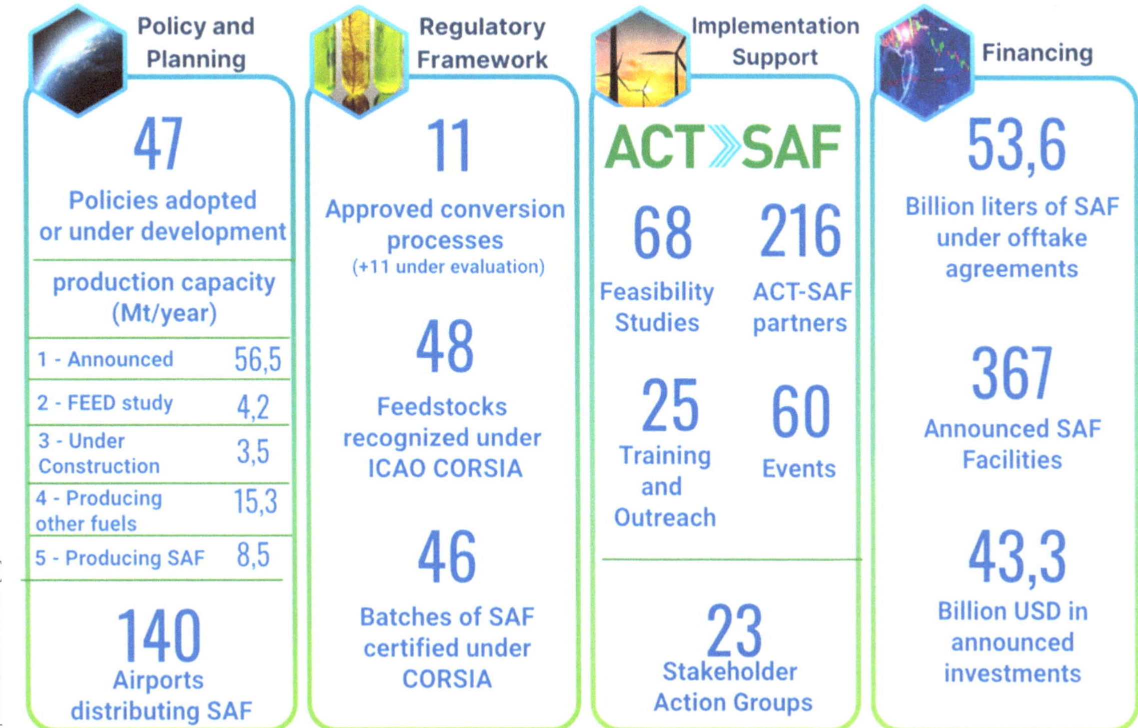

Globally, Europe and North America are leading the efforts to increase SAF utilization and capacity expansion, while the rest of the world is increasingly joining these efforts to develop a global value chain. To support the goals, governments are providing clear incentives for airlines to adopt SAFs, but these vary significantly across regions. The UK, EU, and US SAF usage targets are set out in **Figure 4.3**. Although there is currently little mandate for SAFs in Asia, the Asia Pacific Airlines Association recently called for more government regulations and incentives for SAF use [4.12].

- **UK:** According to the decision taken by the UK Parliament in 2024, the SAF mandate requires 2% use by 2025, 10% by 2030, and 22% by 2040 [4.13]. Continuity of the supply process has been ensured by introducing a purchase price requirement for fuel suppliers who cannot provide supply security. This is considered an offtake agreement. It is usually an agreement to purchase all or a significant portion of the output or product of a specific project. In the case of SAF production, an airline agrees to buy a certain amount of gallons of SAF for a certain period at an agreed fixed price (or a semi-fixed price that varies depending on certain indices) [4.14].
- **EU:** The EU SAF requirement, ReFuelEU, was adopted in October 2023. The requirement will increase over time, reaching 2% by 2025, 6% by 2035, and 70% by 2050 [4.15]. Sweden

and France, in particular, have previously set SAF requirements of 1% in 2021 and 2022, respectively. Norway has had a 0.5% SAF requirement in place since 2020. The EU allows the aviation industry to use a reservation and claim system that separates fuel credit from where it is used until 2035, after which the SAF must be physically supplied at EU airports. They have also introduced financial penalties for fuel suppliers who fail to meet the requirements at airports [4.16].

- **US:** Unlike the UK and the EU, while there is no SAF mandate in the US, there are regulatory incentives and financial incentives to enable the market. Specifically, in the US, the federal government continues to expand laws and policies to incentivize the use of SAFs [4.17]. SAFs are eligible for both the federal Renewable Fuel Standard and various state-level LCFS, and the compliance and credit mechanisms of these programs are seen as ways to help mitigate the impact of higher production costs. In the US, the Sustainable Aviation Fuel Grand Challenge continues as a joint effort by the US Department of Energy, US Department of Transportation, US Department of Agriculture, and other government agencies to expand domestic SAF production, reduce costs, and increase sustainability [4.18]. SAF production is a priority in this effort, with a production target of 3 billion gallons by 2030 and 35 billion gallons by 2050. Improving the entire supply chain, including logistics and blending, is an activity within this program [4.19].

Figure 4.3 UK, EU, and US SAF usage targets [4.12].

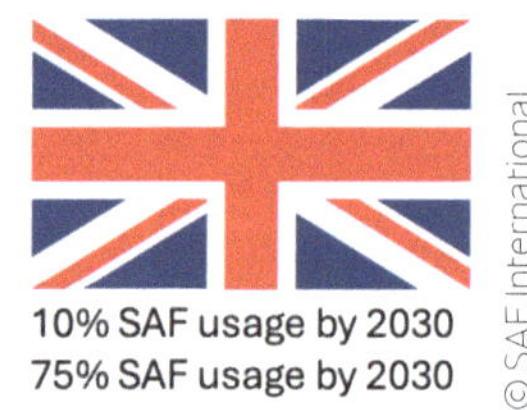

The current regulations have created many offtake opportunities for both airlines and business/private aviation, as the demand for SAF exceeds the supply. Thus, SAF offtake agreements are increasingly being negotiated and executed, and SAF production facilities continue to be financed [4.20]. **Table 4.1** provides information on publicly announced purchase agreements by global and US airlines based on their projections for using SAFs [4.21]. As the demand for SAFs increases, not only will more SAF offtake agreements be signed and more SAF production facilities financed, but lenders will also commit larger amounts to such SAF projects. In addition, an increase in innovation by producers and offtakes to increase SAF supply is anticipated, including efforts to produce "net-zero oil" by investing in direct air capture and other technologies to develop lower-carbon aviation fuels [4.22].

Table 4.1 Global and US SAF offtake agreements [4.21].

Year	Agreement	Parties	Plant	Technology	Capacity
2022	Offtake	Qantas Aemetis Japan Airlines Aemetis Oneworld Aemetis	Aemetis Carbon Zero Plant-Riverbank California	SAF from zero-carbon-intensity electricity and negative-carbon-intensity hydrogen from waste wood and renewable oils, along with CO_2 sequestration	35M gallons of blended fuel over seven years from 2025 90M gallons of blended fuel over seven years from 2024 350M gallons of blended fuel over seven years from 2024
2021		British Airways Phillips 66	Phillips 66 Humber Refinery North Lincolnshire	SAF processed from sustainable waste feedstock	N/A
2021		Etihad Airways Tadweer	First waste-to-SAF plant in the Middle East	SAF processed from municipal commercial and industrial waste	140M gallons of SAF per year
			US		
2021	Offtake	American Airlines Aemetis	Aemetis Carbon Zero Plant-Riverbank California	SAF from zero-carbon-intensity electricity and negative-carbon-intensity hydrogen from waste wood and renewable oils, along with CO_2 sequestration	280M gallons of blended fuel over seven years from 2024
2021		Delta Aemetis			250M gallons of blended fuel over seven years from 2024
2022		JetBlue Aemetis			35M gallons of blended fuel over seven years from 2025
2021		Southwest Airlines	Bayou Fuels Facility-Natchez Mississippi	SAF from sustainable feedstock and renewable power from a neighboring solar facility, as well as carbon capture	219M gallons of preblended fuel over 15 years from 2026
2021		United Airlines Honeywell Alder Fuels	N/A	Conversion of abundant biomass into drop-in replacement crude	1.5b gallons after commercialization in 2025

In general, airlines are becoming increasingly cautious about sourcing and using more SAFs. Regulations and mandates, to which airlines with a global network must adhere, will result in switching to SAFs. Many new and existing fuel suppliers are investing time and resources in identifying new production methods and feedstock sources to meet market needs. The production, transportation, and integration of SAFs will be a key aspect of aviation operations and support SAF utilization in the coming years and decades.

Drop-in Fuels in Terms of Safety and Cost

Drop-in fuels are synthetic and fully interchangeable substitutes for conventional petroleum-derived HCs, meaning they do not require engine or fuel system modification [4.23]. They can be used in existing engines "as is"; they can be mixed with conventional fuels or used even in their pure form. However, the engine should not be used above the recommended rate for modified lubricants, risking loss of warranty, type approval, and modification [4.24]. In many cases, the use of alternative fuels can improve the emission profile by optimizing engine control parameters. According to this definition, SAFs are not yet approved for use in aircraft at 100% concentration, but when mixed with conventional fuel, they can be "dropped" into aircraft engines without any equipment changes or modifications [4.25]. When evaluated according to flight parameters (including the percent share of total industry CO_2), the ease of use and energy density make SAFs attractive drop-in fuels compared to other options [4.26].

While other new technologies, such as hydrogen and electric aircraft, will also play a role in decarbonizing the aviation industry, it is known that such innovations will take time to become viable, especially for medium- and long-haul flights, which account for 73% of the industry's emissions [4.27]. Under this projection, the commercial availability of new technologies will not necessarily translate into widespread deployment. It would also require significant supply, distribution, and infrastructure changes for broader adoption of anything other than a jet fuel equivalent like SAFs. Therefore, SAF being the aviation industry's drop-in fuel would be a valuable investment in the industry. In achieving the net-zero emissions target set for 2050 under current accepted projections, SAF use will become even more critical as a carbon reduction arm for airlines flying to domestic and long-haul destinations. Abiding by this IATA specification makes SAFs easy-to-use drop-in fuels, which can be blended with conventional jet fuel (kerosene). Since SAFs have the same physical and chemical properties as conventional fuel, airports do not need to bother with costly infrastructure changes. The Federal Aviation Administration (FAA) and EASA have accepted this drop-in fuel concept [4.28]. Expanding the use of SAFs has been widely included as a key component of major industry initiatives to reduce aviation emissions. SAF usage is slowly but steadily increasing. The projection determined as the target for the drop-in blending ratio of SAF usage includes a gradual increase, as indicated in **Figure 4.4** [4.29].

Blending: For safety reasons, all SAFs must be blended with traditional Jet A in order to be used in aircraft [4.30]. There needs to be a system and approach for where and when SAFs produced in an independent facility will be blended with Jet A. Fuel quality control requirements require that SAFs from an independent facility be blended with Jet A at a terminal. Airport infrastructures are currently equipped with blending equipment, software, and personnel similar to those that are designed to blend road transport fuels. Thanks to the existing airport infrastructures and oil transportation pipelines around the world, fuel producers and suppliers can send fuel directly to an airport thousands of miles away with the help of pipelines to meet the needs of end users (airlines). Regulations state that clean SAFs from two different routes cannot be mixed in the same tank for blending with Jet A. There are two methods (**Figure 4.5**) for storing and blending SAF for delivery to the end-user point, and the method selected is determined by fuel handling requirements, terminal operations, and fuel supplier preferences [4.31].

Figure 4.4 Drop-in blending ratio of SAFs [4.29].

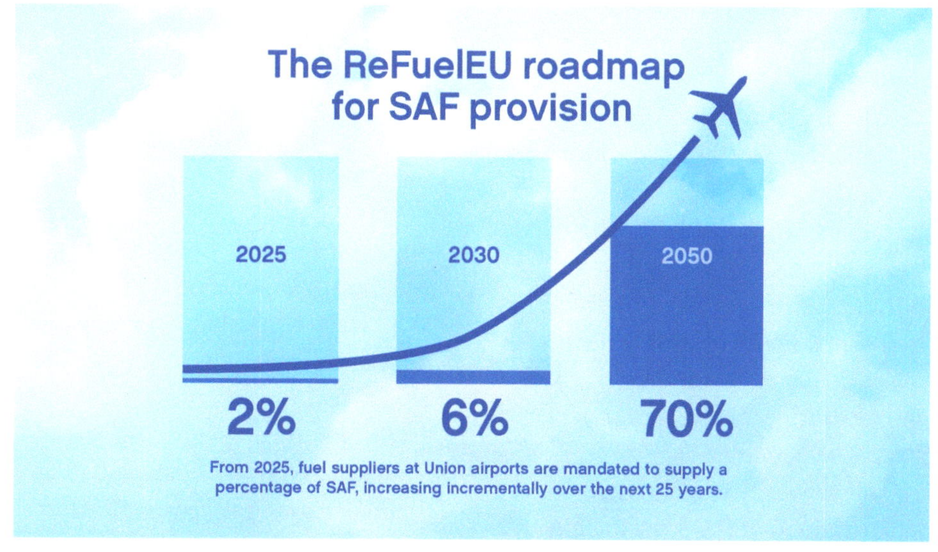

© Neste.

Figure 4.5 Drop-in blending options (a and b) for Jet A and SAFs [4.31].

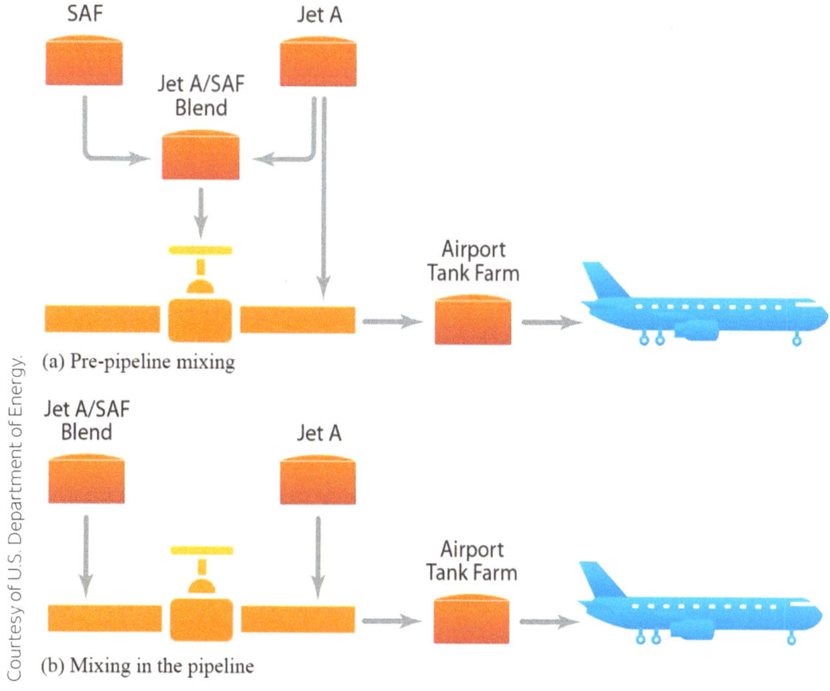

Courtesy of U.S. Department of Energy.

Commercial flights are currently permitted to fly with a blend of SAF and conventional fossil-based kerosene of up to 50%, to ensure compatibility with aircraft, engines, and fueling systems [4.32]. The actual amount of blending allowed varies depending on the production method, but with HEFA and FT, the upper limit is 50%, while for AtJ, it is limited to 30%. The current practice stores SAF and Jet A in separate tanks, and both are tested for compliance with ASTM D7566 and ASTM D1655, respectively. SAF and Jet A are delivered in a specified line at the desired ratio to a third mixing tank where the fuel is tested and certified to ASTM D1655. A second option, not currently implemented with the tier update under ASTM D1655, is to deliver the SAF to a Jet A tank in a similar manner, where the SAF is mixed, tested, and certified according to ASTM D1655. This method is not used because certification and potential fuel-related safety issues at the end-user location have caused fuel tank cleaning and fuel supply delays. SAF/Jet A blends can be delivered to airports via existing pipelines [4.33, 4.34]. Using the same pipelines and trucks to deliver SAF/Jet A blends at airports supports SAF usage and does not increase total costs by ensuring that existing procedures are followed.

- **Safety:** Safety is the most important prerequisite for a new fuel to be used in an aircraft engine from every perspective. Safety issues in aviation fuels include production, storage, transportation, and operation [4.35]. Especially in civil aviation transportation, ensuring the safety of aircraft and aircraft engines is the most important feature. It is unacceptable to directly adopt any new technology at the expense of aircraft engine safety. Therefore, the aviation industry has developed airworthiness standards and safety assessment methods to ensure aircraft engine safety. The security components that need to be considered due to the use of SAFs are indicated in **Figure 4.6** [4.36].

Figure 4.6 Aviation in terms of safety [4.36].

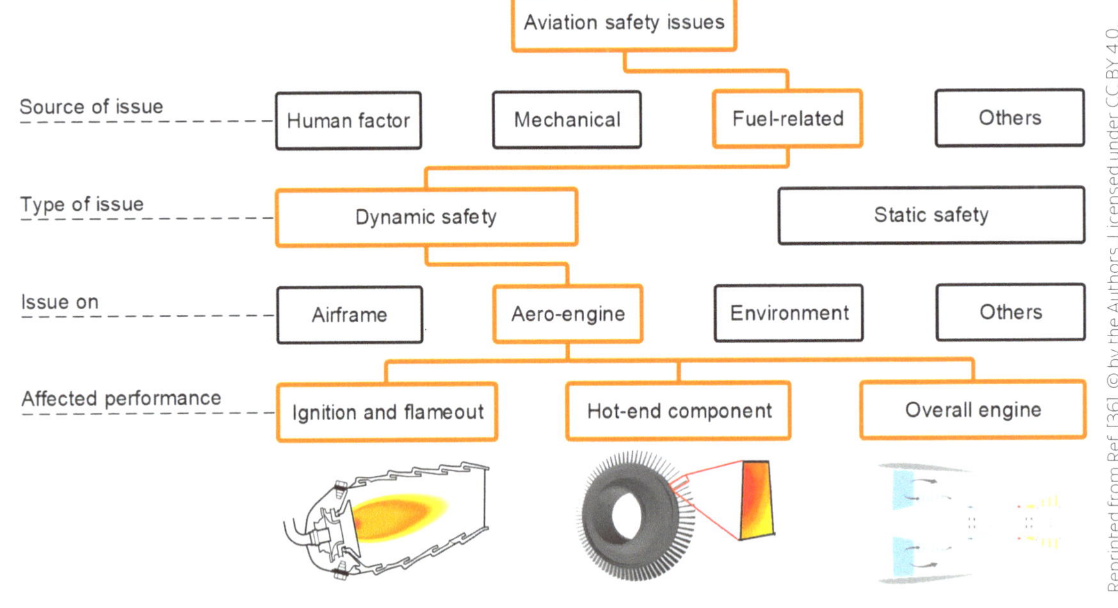

One of the most important components is fuel-related safety. A manufactured fuel specifically designed for aircraft engines must meet both airworthiness and technical standards. The detailed parameters shown in **Figure 4.7** present the major issues to be considered in the effects of SAFs on aircraft engine safety [4.37]. Fuel safety criteria include stages such as certification, fuel transportation in the system, inclusion in the combustion process for power generation, and the desired power functions of the engine.

Figure 4.7 SAFs in terms of safety [4.37].

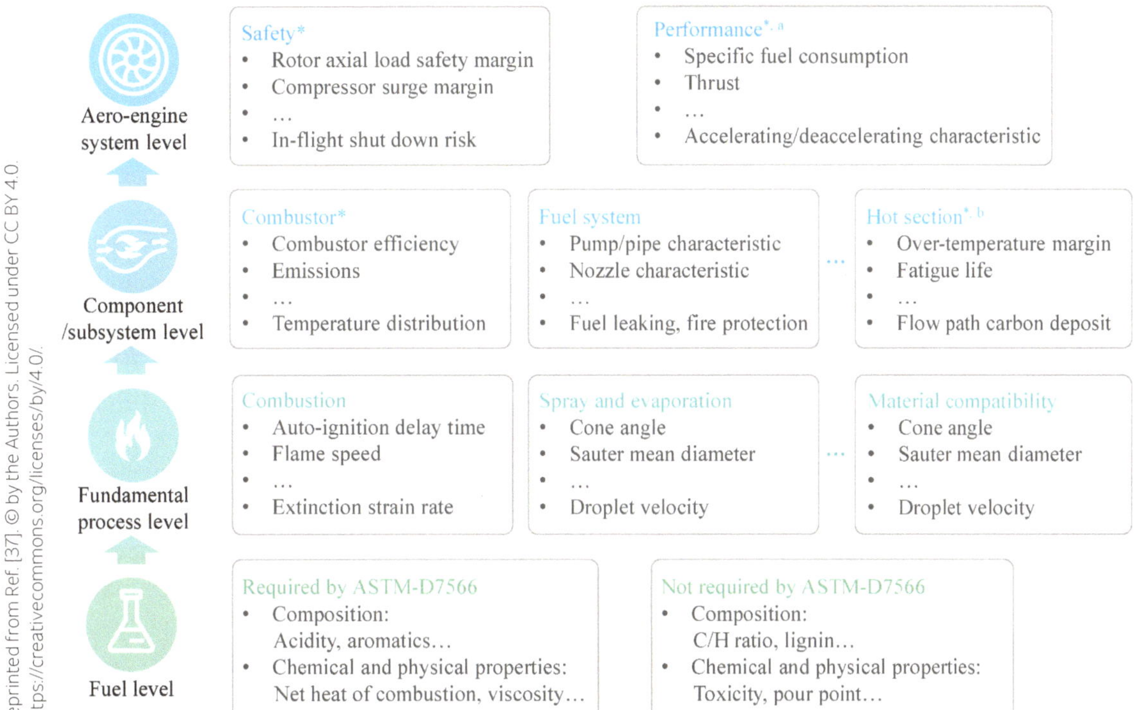

There is very limited published research discussing the safety effects of SAFs at the fuel system and engine level. Even with conventional jet fuels, there can be flight accidents or failures directly related to the fuel. Safety in the use of SAFs is fundamental to their improvement, certification, and implementation. SAFs can affect aircraft engine ignition, flameout performance, and overall aircraft engine performance. In recent years, various national and international organizations have developed standards related to SAFs. To ensure that there are no safety problems in the application of SAFs to aircraft engines, airworthiness standards developed by airworthiness authorities require provisions for verifying the compatibility of aircraft engines after the application of new fuel [4.38]. As a specific component of aircraft engines, SAFs must also comply with ASTM technical standards governing fuel characteristics. Airworthiness and technical standards together provide guidelines for SAFs. In practice, the standards intersect in their scope of application and refer to each other, forming the SAF-related standards system. However, due to the differences in the purposes of airworthiness and technical standards, there are significant deviations in the safety requirements of SAFs [4.39]. The characteristics, advantages, and disadvantages of the two types of standards and their relationships with each other should be clarified to provide a basis for further improvements in the SAF-related standards system. In particular, the safety effects of SAFs that have been certified and determined to be suitable for use during the acceleration and deceleration processes of an engine, such as LTO, must be defined.

Cost: Increasing the raw material supply and developing practical production technologies are important factors in determining the SAF usage rate and cost. SAF technology is advancing rapidly, and significant scaling is expected by 2030 [4.40]. The industry and governments are working toward commercial aircraft being permitted to fly on 100% SAF in the near future. However, airline adoption of SAFs has so far been limited by the lack of production capacity and cost premium [4.41]. Fossil-based Jet A costs about $1000 per ton, and SAFs typically cost between 3 and 6 times that amount. This presents a significant challenge in an industry particularly sensitive to operating costs, especially fuel production. Airlines are very price-sensitive because jet fuel accounts for approximately 30–40% of operating costs [4.42]. One of the cost drivers for using SAFs is the ability to blend and transport to the aircraft as quickly as possible. The lowest cost and fastest timeline scenario would be to use existing tanks at a terminal. Fuel industry stakeholders agree that a significant increase in production capacity, as well as strong government support, will be required to offset the higher cost of SAFs [4.40, 4.41]. The price of SAFs varies greatly depending on various factors such as raw material availability, production technology, government incentives, and market demand [4.43]. Corporate buyers only pay the price premium of SAFs, while airlines pay the cost equivalent to conventional jet fuel. Instead of receiving a predetermined cost, companies always ask their transaction partners for a clear price breakdown of SAFs compared to conventional jet fuel. The SAF price premium is derived as follows: SAF cost minus raw material price, production and logistics costs, conventional kerosene price, including sustainability certification, and any government subsidies [4.42, 4.43]. Significant steps have been taken to reduce costs in jurisdictions such as the EU, US, and UK, all at different stages and with different approaches [4.44]. Given the higher cost of SAFs compared to standard jet fuel, businesses can play a leading role by integrating SAFs into their OPEX, thereby increasing the demand and potentially reducing costs over time. As the production scales up and the availability of various feedstocks increase, prices are expected to fall, eventually closing the gap between the costs of SAF and fossil-based jet fuel. Developing the SAF industry and thus closing the gap will require collaboration between governments and industry leaders to implement synchronized policies, provide incentives, and facilitate the investments needed to scale up SAF production. The US, which has been a pioneer in using SAFs, has both federal and state support. **Figure 4.8** shows the federal incentives for Renewable Diesel (RD) and SAF for California (CA), Oregon (OR), and Washington (WA), and the combined values of federal and state incentives [4.45].

Figure 4.8 Federal combined incentive values for SAF use [4.45].

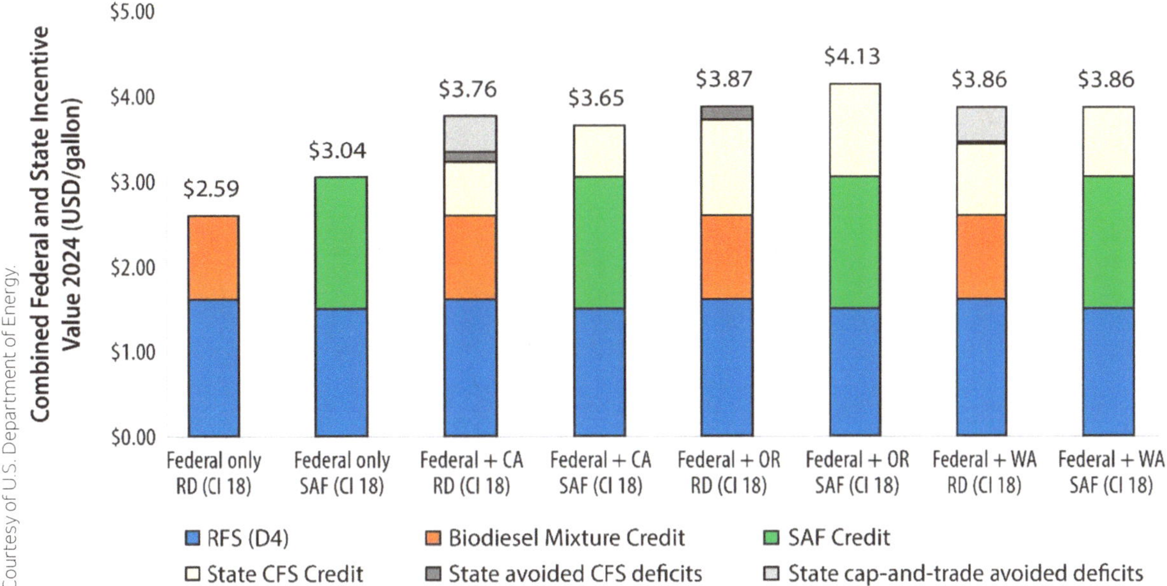

The Inflation Reduction Act of 2022 enacted Section 40B, which provides a SAF-specific tax credit of up to $1.75 per gallon for both domestic production and imports between December 31, 2022, and January 1, 2025. In 2025, the Inflation Reduction Act's 45Z Clean Fuel Production Credit (CFPC) replaced 40B, which provided up to $1.75 per gallon for SAFs that met various requirements. Unlike 40B, the 45Z tax credit does not apply to imported SAFs. There is also the FAA $244.5 million Fuels for Sustainable Aviation Transition (FAST) Grant Program to support SAF production, transportation, storage, and blending [4.46]. The EU, by contrast, takes more of a "stick" approach, with a SAF mandate known as ReFuelEU that kicks in in 2025 requiring 2% of all jet fuel to be sustainable and increasing from there. A comparison of the production cost along conversion pathways compared to the wholesale cost of petroleum-derived jet fuel for investing in a particular fuel conversion technology within the EU is shown in **Figure 4.9** [4.47].

Figure 4.9 Comparison of the production costs of SAF types with respect to jet fuel [4.47].

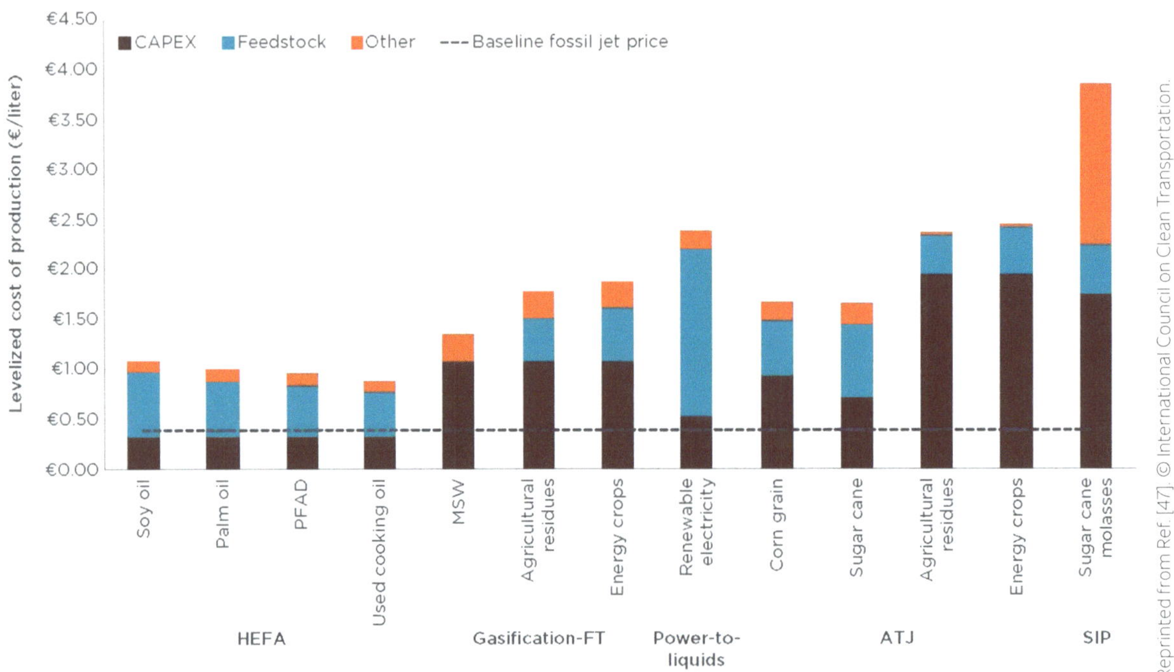

Overall, the HEFA route appears to be the cheapest source of SAF per liter, with a levelized cost of around $0.91 to $1.13 per liter depending on the raw material. Due to the relatively low CAPEX of the HEFA conversion process, most of its cost is derived from the feedstocks [4.48]. A key factor in reducing SAF costs is addressing the logistics of raw material supply and supply chains. Different regions offer different carbon intensities (CIs) and access to raw materials, and optimizing these supply chains can help reduce the financial burden of SAF production. However, this optimization is currently largely dependent on government incentives, as SAF projects may struggle to achieve profitability without proper regulatory support. As the global SAF industry matures, these supply chains should become more robust and provide an additional avenue for reducing SAF costs. While legislation is crucial to driving SAF production and adoption, it should be aligned with the ultimate goal of reducing CI. Companies can achieve lower costs and create more sustainable processes by producing SAFs as efficiently as possible and working across supply chains to reduce emissions [4.49]. The increasing demand for SAFs can support the improvement in production processes and gradually reduce costs. Businesses have the opportunity to contribute to the scale-up of SAFs by leading business flights where necessary. However, without consistent and cohesive government support, using SAF risks remaining a fragmented and expensive solution for airlines.

Demand and Availability

Global aviation demand is expected to more than quadruple by 2050, requiring an average reduction of around 1.4 billion tons per year to achieve net-zero CO_2 emissions [4.50]. Supported by policymakers worldwide, SAFs will meet the ever-increasing demand to reduce emissions. The US and the EU are leading the development of sustainable economic and environmental policies in the aviation sector. In this context, the management of the demand for SAFs is largely dependent on government incentives and is a lively managed process. While the EU and the US have the largest proportion in the efforts to direct the use and capacity expansion of SAFs by managing the process, the rest of the world is increasingly trying to join these efforts to develop a global value chain [4.51].

In addition to increasing aviation demand, decarbonization efforts, and other significant works by SAF producers and refiners, increasing the supply, techno-economic developments, and feedstock supply will drive growth in the global SAF demand. The quantities currently produced worldwide are low (less than 1% of total jet fuel demand), but with coordinated support, including effective policy frameworks, these quantities can be increased significantly [4.52]. Many major refiners and oil and gas producers are involved in SAF production and technology development. Major producers are partnering with SAF producers to cost-effectively scale up the capacity. Production and demand in 2040 and 2050 appear to be difficult to predict due to inadequate public announcements from fuel manufacturers [4.53]. Based on regulations, commitments, and capacity announcements, the global SAF demand is expected to grow by approximately 29% annually through 2050. **Figure 4.10** shows projected global SAF demand growth [4.54].

Figure 4.10 Estimated global SAF demand through 2050 in million barrels per day [4.54].

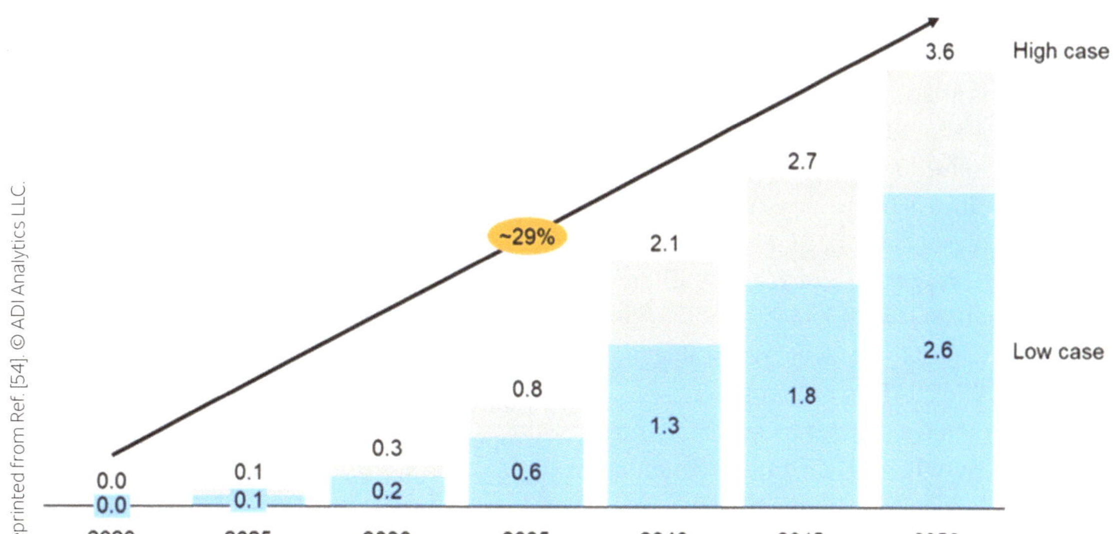

Several airlines have set commitments that could potentially increase the SAF demand. Collaborations between airlines, fuel manufacturers, and technology providers create opportunities for the SAF market. Joint ventures and partnerships leverage the expertise and resources of multiple organizations to accelerate SAF production, improve distribution networks, and address infrastructure challenges. Many fuel industry companies have announced plans to enter the SAF market by 2030 [4.55]. Companies currently engaged in SAF production can be categorized according to the technologies they employ. In addition, existing agreements to purchase SAF highlight that most airlines have established partnerships primarily with biofuel producers [4.56]. Considering the partnerships and production costs, more than 60% of SAF demand is projected to be met by HEFA fuels in 2030 [4.55, 4.57].

As a result of the efforts to meet SAF demand, production volumes reached 1 million tons (1.3 billion liters) in 2024, which is twice the 0.5 million tons (600 million liters) produced in 2023. SAFs accounted for 0.3% of global jet fuel production and 11% of global renewable fuel in 2023 [4.58]. This was significantly below the previous estimates of 1.5 million tons (1.9 billion liters) of SAF production in 2024, and major SAF production facilities in the US have postponed production increases until the first half of 2025. SAF production is expected to reach 2.1 million tons (2.7 billion liters) in 2025, or 0.7% of total jet fuel production and 13% of global renewable fuel capacity [4.57, 4.58]. SAF investments are increasing, but at a slower pace compared to the targeted projection. To meet the future demand, governments should use their potential to accelerate progress by reducing fossil fuel production subsidies and replacing them with strategic production incentives and clear policies that support a future built on renewable energies, including SAFs.

- **Availability:** Due to country-specific policies and economic reasons, SAFs are not yet available at all airports worldwide. With current technological methods and available raw materials, the amount of SAF produced in a given region limits the increase in its use. Only about 0.1% (=240,000 tons) of liquid fuel needed worldwide is produced from sustainable sources [4.59]. Although this is not yet sufficient to use large quantities in flight operations at present, the SAF sector has gained significant momentum with regulations and policies adopted. The lack of availability can also be partially attributed to the limited number of locations where it is produced [4.60]. However, common policies established through intergovernmental regulations have made it necessary for countries around the world to keep up with the changes in air transportation.

Considering all attempts to use SAFs, **Figure 4.11** shows the growing interest in its use, including all research activities on SAFs worldwide [4.61]. This map visually represents where news and activities published in the ICAO Global Aviation Alternative Fuels Framework (GFAAF) have occurred since 2009. Airlines are implementing regulations in accordance with the accepted policies and systematically approving SAF purchases. In this process, a significant number of purchase agreements are being made for both airlines and commercial/private aviation although the SAF demand exceeds the supply. European aviation meets the same stringent quality standards as the US, and partnerships are being formed to supply SAFs. **Figure 4.12** shows the initiatives and projects to use the SAFs published in the ICAO GFAAF [4.62].

Figure 4.11 The world framework of SAF activities [4.61].

Reprinted from Ref. [61].

Figure 4.12 Initiatives and projects to use SAFs [4.62].

Reprinted from Ref. [62].

Research on SAF production, regional feedstock sources, available technologies, and supporting initiatives has been conducted, and many more are underway. More than 100 publicly announced SAF projects have been registered through 2022, with hundreds more to be registered through 2050 [4.63]. Over the past ten years, approximately 190 companies have announced their intention to produce sustainable fuels at 330 locations worldwide. Out of those 190 companies, only 18 are currently producing SAFs, and only a few of them are producing on a large scale. The US is leading the way in terms of national and state incentives. This effort will greatly support the availability of SAFs. **Figures 4.13** and **4.14** show the ongoing deliveries and locations of airports receiving SAF batches worldwide [4.64, 4.65].

Figure 4.13 Airport use of SAF_1 [4.64].

Reprinted from Ref. [64].

Figure 4.14 Airport use of SAF [4.65].

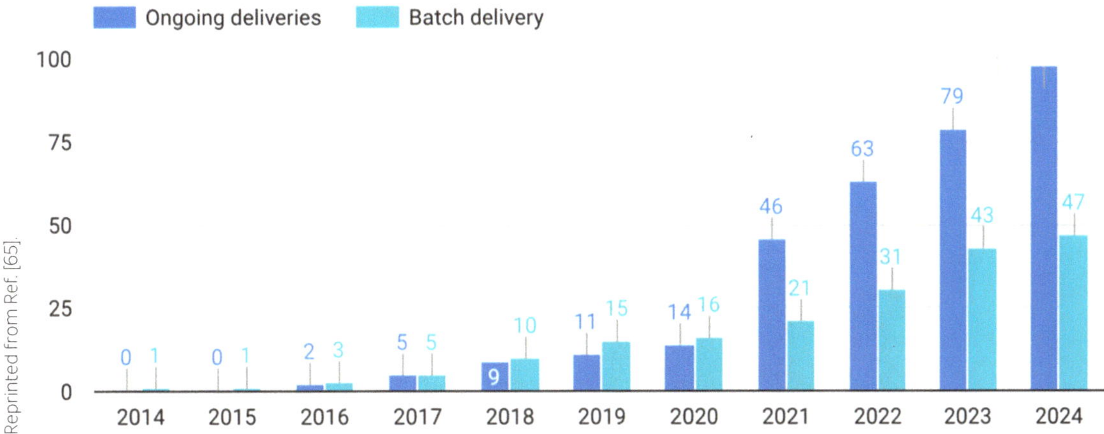

Current projections suggest that SAFs will eventually be required at all airports, but by 2034, fuel producers will have the opportunity to supply aviation fuel with a higher percentage of SAFs to certain airports due to low or no SAF availability at other airports. Many airports now offer SAFs on a continuous or bulk basis. More details on these SAF distribution initiatives are presented elsewhere in this chapter. "Batch delivery" refers to airports that receive single-time SAF deliveries, typically associated with a single flight or series of flights. "Ongoing deliveries" refers to airports that advertise the continuous availability of SAFs for their operations. As shown in **Figure 4.14**, there was ongoing delivery of SAFs at 94 airports in 2024.

The SAF market in the US is witnessing significant growth due to the strong emphasis on environmental sustainability and the presence of several major airlines [4.66]. There is a high demand for SAFs, driven primarily by regulatory initiatives and corporate sustainability goals. The region has a well-established supply chain with several SAF-manufacturing facilities and refineries, as shown in **Figure 4.15** [4.31].

Airports do not purchase jet fuel, but they play a key role in preparing their locations for SAF/Jet A blends. **Figure 4.15** shows the current US jet fuel pipeline [4.31]. Consumer preference for eco-friendly air travel is also increasing, leading to increased adoption of SAFs. Using the existing jet fuel pipeline to transport SAF will help meet industry goals. Airports are bringing together airlines, fuel producers, suppliers, terminals, and other stakeholders in the fuel supply chain to enable SAF purchasing and blending at their geographic location and scale up for future larger volumes. The EU has become a significant market for accessible SAFs, with many countries complying with regulations and policies [4.67]. The region has a strong supply chain with multiple SAF production facilities. Consumer preference for sustainable travel options and European airlines' commitment to carbon reduction are further fueling the growth of the SAF access market in the EU.

Figure 4.15 Major US pipelines carrying jet fuel [4.31].

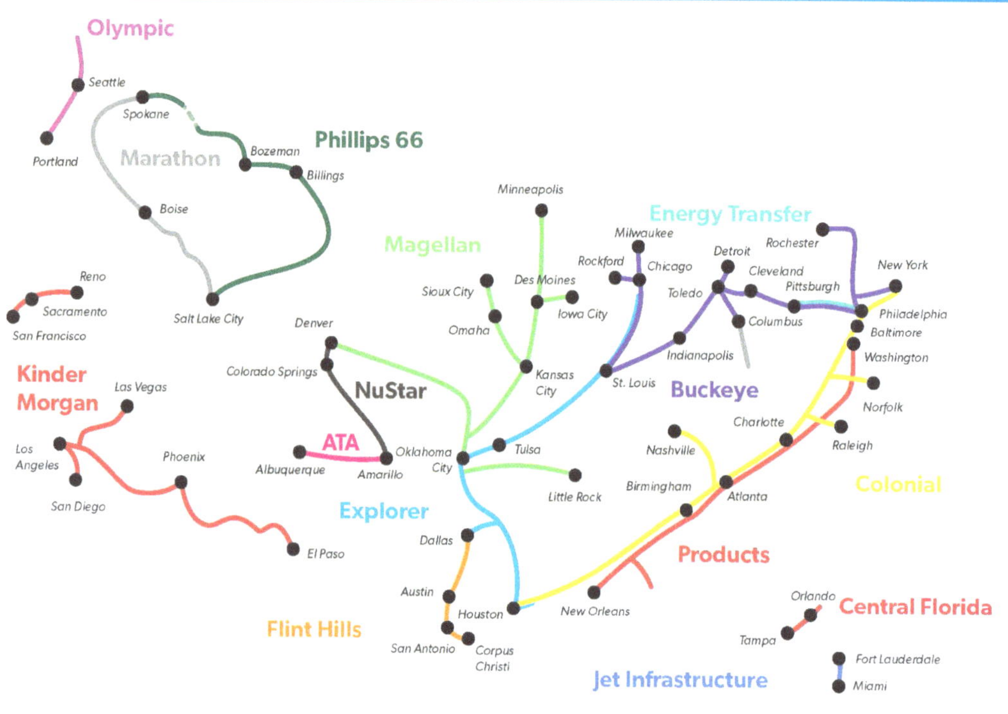

Additionally, the SAF market in the APAC region is experiencing significant growth due to increased air traffic and a focus on sustainable development. Countries such as Australia, Japan, and Singapore are investing in SAF production and infrastructure development [4.62]. As in other continents, increasing consumer awareness and preferences for sustainable air travel and government support have increased the demand for SAFs in the region. However, the market is still in its infancy and must overcome challenges such as limited production capacity and supply chain access.

Specific Fuel Quality Requirements and Fuel Systems

Aviation fuels are a complex mixture of HCs as their properties vary depending on the production source and process. Aviation fuel quality assurance is based on certification at the point of production and procedures to verify that the quality of the fuel in question remains within the defined limits of the standard and does not change significantly during distribution and delivery to aircraft [4.68]. Two basic quality assurance concepts, known as batches and traceability, ensure that aviation fuels are used at the same level of safety at each distribution point [4.69]. A fuel batch is defined as a specific quantity of jet fuel that can be characterized by a series of test results. Therefore, refiners can ensure that the batches are homogeneous and thus the test results are representative of the product supplied. Traceability for aviation turbine fuel is defined as the ability to trace different fuel batches back to the original point of production through the distribution system using accurate documentation. The FAA sets "airworthiness standards" for the design and operation of aircraft. Similarly, fuel is considered an operational limitation for aircraft and engines, and a safety standard is expected. The aviation industry recognizes that alternative fuels should be "dropped in" to commercial engines, pipelines, fuel farms, and all other distribution and storage channels without modification, thus requiring no new equipment or infrastructure [4.70]. The industry has developed rigorous testing requirements to determine whether a fuel can be considered "drop-in" and to compare it to petroleum-derived jet fuel. Jet fuel quality is carefully controlled through multiple harmonized standards adopted worldwide. Fuels that comply with a new sustainability certification program approved by CORSIA in the aviation sector can be used safely in aircraft [4.71]. Like conventional jet fuel, SAFs require rigorous technical certification. SAFs are certified by sustainability certification schemes in accordance with standards set at global and EU levels under the ICAO CORSIA framework and the RED [4.70, 4.71]. This process, along with the development and management of specifications for alternative aviation fuels, is carried out in the US by ASTM International. The United Kingdom Ministry of Defence maintains another jet fuel quality standard, Defense Standard 91-091 (Def Stan 91-091-Turbine Fuel, Kerosene Type, Jet A-1), which is widely used outside the US [4.72]. ASTM International is a consensus standards organization comprised of aviation industry experts, including aircraft and engine manufacturers, fuel system equipment manufacturers, fuel manufacturers, suppliers, users, and other interested parties. **Figure 4.16** shows the standards that determine fuel quality in the aviation industry [4.36].

Figure 4.16 Quality standards of aviation fuels [4.36].

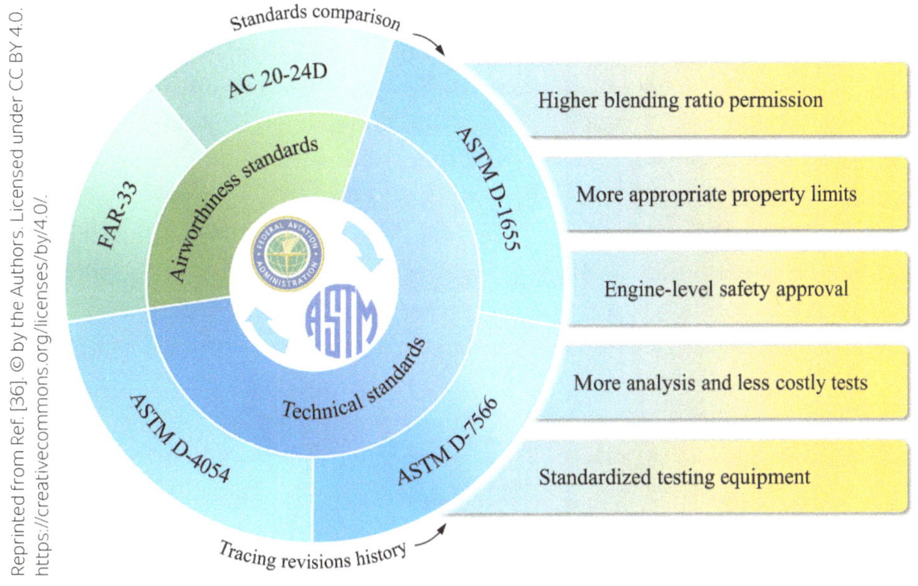

Aircraft and engine manufacturers worldwide currently use ASTM D1655 (Standard Specification for Aviation Turbine Fuels) specifications to define the quality and conformity of their jet fuel products.

The effects of the chemical components that make up the fuel on fuel properties are effective in determining the standards. In particular, the parameters affected by fuel composition shown in **Table 4.2** (performance and operability characteristics) play a major role in determining the limits of the standards in terms of how well the fuel works and whether the fuel will work safely [4.73].

Table 4.2 Effect of jet fuel composition on performance and operability [4.73, 4.76, 4.77].

Parameters	Properties	n-Alkanes	Isoalkanes (weakly branched)	Isoalkanes (strongly branched)	Cycloalkanes Monocyclic	Cycloalkanes Fused bicyclic	Aromatics
Performance	Specific energy	+	+	+	+	NA	–
	Energy density	–	–	–	+	+	+
	Thermal stability	+	+	+	+	+	NA
	Emission/sooting	+	+	+	+	+	–
Operability	Derived cetane number	+	+	–	NA	NA	–
	Density	–	–	–	+	+	+
	Freeze point	–	±	+	+	+	+
	Emission/sooting	+	+	+	+	+	–

Table 4.3 also lists the basic jet fuel properties and drop-in requirements in ASTM D1655, their effects, and definitions [4.73]. This specification defines the minimum requirements for Jet A and Jet A-1 aviation turbine fuels. ASTM D1655 is intended for civilian applications but can be adapted for military, government, or other specialty uses. Jet A is the fuel type used in the US, and Jet A-1 is used in the rest of the world. These two fuels are nearly identical, with the primary difference being the freezing point, which is −40°C for Jet A and −47°C for Jet A-1 [4.74, 4.75]. The procedure and rules for getting approval for a new fuel from engine and aircraft manufacturers are described in ASTM D4054 (Standard Practice for Evaluation of New Aviation Turbine Fuels and Fuel Additives) [4.76]. The new aviation fuel must pass ASTM's D4054 evaluation process to determine whether it is equivalent to conventional jet fuel. As shown in **Figure 4.17**, the physical analysis and engine testing performed by the manufacturers under ASTM D4054 are extensive and occur over a long period of time. Meeting this standard involves four parts. The first part is a six-month review of the fuel's specifications and characteristics (called Tier 1 and Tier 2); then, if it is found to be fit for purpose—requiring a shipping cost of around $50,000 (in 2024) and 200 L of clean product—it is then tested for six months with aviation partners such as Airbus, Safran, and Rolls-Royce at a cost of around $350,000 (in 2024). Positive results will continue for two to three years of further testing (Tier 3 and 4) with a large price tag and a minimum of 100,000 L of pure fuel. If successful, this will lead to FAA approval and a vote by ASTM's expert panel, which must be agreed upon by all participants before the fuel is released for blend [4.77].

Table 4.3 Specific jet fuel properties and drop-in requirements [4.73, 4.77].

Properties	ASTM units	Effect	Description
Specific energy	**MJ/kg**	Combustion	Provides fuel efficiency by reducing takeoff weight, which is critical for mass-limited missions
Energy density	**MJ/L**		The most important criterion for volume-limited missions or military operations involving refueling
Surface tension	**mN/m**		High values may prevent spray fragmentation and atomization
Thermal stability	**mmHg/scale**		Determines the ability of the fuel to maintain its properties at high temperatures in fuel system components
Derived cetane number	**Digit**		Important for blowout limit stability in air deficiency (lean)
Aromatic concentration	**vol. %**	Composition	Combustion of highly aromatic fuels produces smoke and carbon/soot deposits, but seals must be properly inflated to prevent leaks
Density	**kg/m³**	Volatility	Used to calibrate measuring equipment and determine aircraft payload weight and range
Flash point	**°C**		Maximum temperature for fuel handling and storage without serious fire hazard
Distillate temperature	**°C**		Fuel volatility and ease of vaporization at different temperatures are determined by distillation; this affects vapor and entrainment losses, vapor lock, flammability hazards, and engine-starting characteristics
Viscosity	**mm²/s**	Fluidity	Viscosity measures the resistance of the fuel to flow; lower temperatures create greater resistance
Freezing point	**°C**		The temperature at which flow through filter screens into the engine may be restricted due to the formation of wax crystals

Figure 4.17 ASTM D4054 evaluation process for new aviation fuel [4.78].

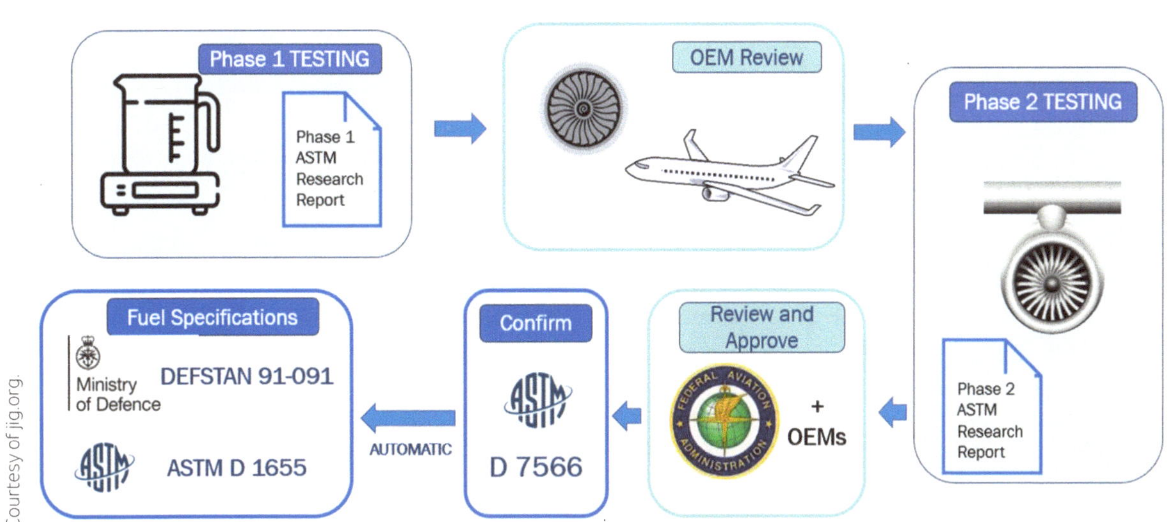

ASTM D7566 amendments ensure that any accepted product will behave equivalently to conventional jet fuel, and fuels that meet ASTM D7566 can later be recertified as meeting one of the conventional jet fuel standards, such as ASTM D1655 or Def Stan 91-091 [4.78]. Since the first biofuel test flight on a commercial aircraft in 2008, a huge amount of work has been done by the industry and its partners to certify SAFs. ASTM International has enabled approximately three-quarters of a million flights using SAF/conventional fuel blends since 2011. Once the new fuel is approved to this standard, the new fuel or fuel blend component is certified to ASTM D7566 (Standard Specification for Aviation Turbine Fuel Containing Synthesized Hydrocarbons) [4.79]. SAFs certified under ASTM D7566 can be blended with conventional jet fuel and used in commercial aviation. A D7566-certified SAF is blended with conventional jet fuel up to its maximum allowed blend ratio. SAFs are typically certified for blending up to 50% with conventional jet fuel. Since pure SAF from two different production methods cannot be mixed in the same tank for blending with Jet A, the various forms of SAF must meet ASTM requirements prior to blending with Jet A/A-1 and also for the final blend. As shown in **Figure 4.18**, after these conditions are met, the blended fuel can be redefined in accordance with the ASTM D1655 conventional jet fuel standard and can be transported through pipelines and used (drop-in) in aircraft [4.80]. Once a fuel has been fully certified, it is recognized as jet fuel and can be used without any restrictions.

Figure 4.18 SAF certification procedure [4.80].

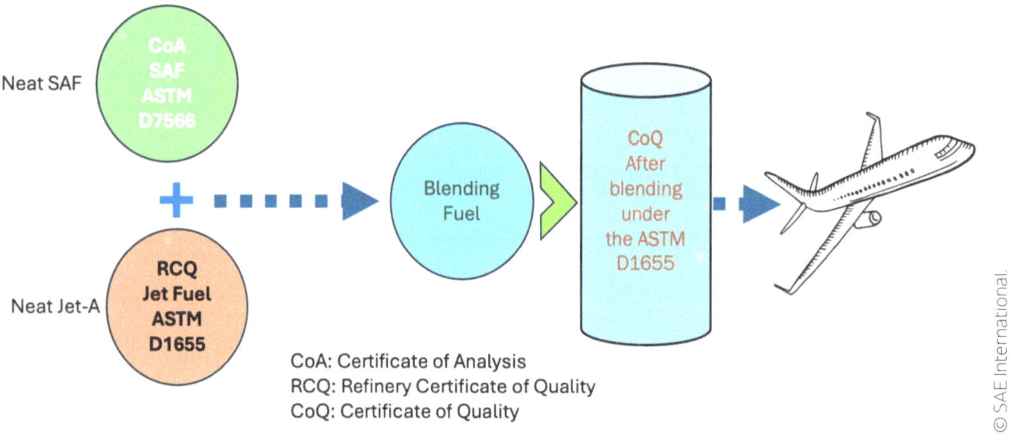

In order to obtain a refinery certificate of quality (RCQ), every batch of petroleum jet fuel produced at the refinery is assigned a batch number and goes through a comprehensive compliance test. As jet fuel passes through the supply chain, a certificate of analysis (COA) is created for each batch, necessitating the retesting of important fuel characteristics. SAF facilities produce a quality certificate attesting to adherence to the relevant ASTM D7566 annex [4.81]. A COA is also produced at the point where Jet A and SAF are combined, as well as for every movement along the supply chain, because a blended fuel would pass through several terminals. This blending procedure ensures seamless integration into the current fuel supply chain and helps mitigate potential supply constraints. Thus, SAF coprocessed at an oil refinery would be ASTM D1655-certified and would travel by pipeline to terminals and from there by pipeline to the airport via trucks, depending on the airport's fueling infrastructure. In 2009, ASTM D7566 was amended to allow FT fuels to be made from biomass, and since then, improvements have been made chronologically to maintain and improve quality, as shown in **Figure 4.19** [4.36].

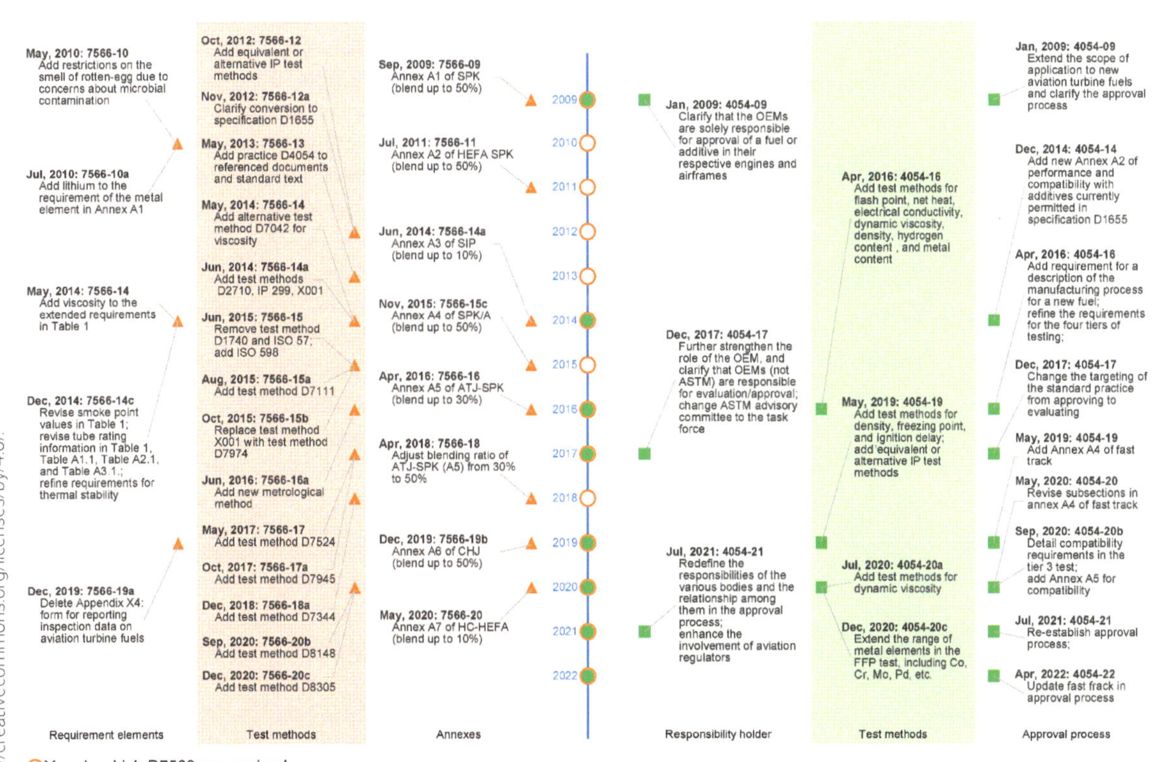

Figure 4.19 ASTM standard improvements for SAFs [4.36].

More SAF-specific specification elements may appear in future standards. The specification elements in existing SAF-related standards have evolved from ASTM D1655. With the increasing diversity of SAF processes and raw materials, new specifications will be required in the standard. This trend is evident from the historical revisions of ASTM D7566 in **Figure 4.19**. The blend percentages given are the maximum allowable values; however, ASTM D7566 fuel specification requirements may limit the blend ratio for a given batch of synthetic blend components to a lower level. Airports do not purchase jet fuel, but they play a key role in preparing their locations for SAF/Jet A blend use [4.82]. Commercial aircraft manufacturers certify that their aircraft will meet the required performance and operational requirements when operating on fuel of a quality that meets certain criteria defined in an internationally recognized specification. If an operator uses a fuel that meets this specification, the aircraft will operate as expected. There is potential for 100% SAF in the future, but there is no effective infrastructure to achieve this, yet. The reason for this is that the fuel system and combustion performance also require a blend that contains at least 8% aromatics within the Jet A aromatic content range.

- **Fuel systems:** An aircraft fuel system plays a key role in loading, storing, managing, and delivering fuel to the propulsion system, which includes an aircraft's engine. The aircraft fuel system consists of precision system elements that direct the continuous flow of clean fuel from the fuel tank to the aircraft engine [4.83]. Aircraft fuel systems encompass a wide range of components, including fuel tanks, pumps, valves, filters, pipes, gauges, and delivery systems. There are different systems according to their types (gravity feed fuel system, fuel injection system, pump feed system) and their components (fuel tank, fuel valve, fuel pump, fuel filter, fuel gauge, fuel line). Aircraft fuel systems are of various types depending on the complexity and size of the aircraft. The most commonly used aircraft fuel systems are pump feed system, fuel injection system, and gravity feed fuel system [4.84]. These components work together to store the fuel, transfer it from the tanks to the engines, monitor fuel quantity and quality, and protect against potential hazards such as fuel leaks and vapor explosions. One of the most important issues in the fuel system until it is included in the combustion in the engine is that the system cuts off the fuel for safety and preserves the existing fuel.

Ensuring the safety of the fuel tank in the aircraft fuel system is a basic requirement for flight. In commercial and military aircraft, fuel tanks serve not only to store fuel, but also as a thermal sink to dissipate heat generated from the engine. Once fully charged and in use, there is a space above the fuel called a void that contains vaporized fuel mixed with air, which contains the oxygen needed for combustion. Under the right conditions, this mixture can ignite. Fuel vapor cannot burn in air with less than 11% oxygen content [4.85]. Fuel tank inerting systems increase aircraft safety by reducing oxygen in the fuel tank to prevent combustion even when a spark is present. In order to prevent accidental combustion of the fuel, nitrogen-enriched air (NEA) is fed to reduce the oxygen content in the tanks and inerted. The fuel–air medium consists of steam, oxygen, nitrogen, and Jet A fuel. If a certified and blended SAF is used, safety can be ensured in situations that may occur in the fuel system [4.86]. In fuel safety tests conducted, all safety criteria results of SAF-blended fuels in the fuel systems of aircraft before flight, during flight, and landing were approved.

The aircraft industry is constantly seeking innovative and sustainable solutions to reduce GHG emissions and mitigate aviation's environmental impact.

In terms of materials and safety in the fuel system, the use of SAFs provides full compatibility criteria when ASTM D1655 or Def-Stan 91-091 is used as the basis. There are no data in the current literature and industry from SAF manufacturers and leading institutions and organizations that SAFs are incompatible with current aircraft engine technology and fuel systems [4.87]. Additionally, the criteria set by ASTM standards include clear limitations that include features that are compatible with current fuel system technologies. For example, the most obvious application of this is the percentage of aromatic components in the fuel that affect the sealing performance of fuel system components. SAF has properties that do not pose a risk to flight safety, especially in the fuel that reaches the engine from the aircraft fuel tank and enters the combustion process. As a result, SAFs, when used in a blended ratio and compatible with fuel systems found on today's aircraft, offer environmental benefits by reducing carbon emissions, reducing combustion performance, and minimizing pollutants. Additionally, advances in SAF storage, distribution, and safety technologies make fuel systems containing SAFs more applicable and reliable.

References

4.1. Urban, F., Nurdiawati, A., Harahap, F., and Morozovska, K., "Decarbonizing Maritime Shipping and Aviation: Disruption, Regime Resistance and Breaking through Carbon Lock-In and Path Dependency in Hard-to-Abate Transport Sectors," *Environmental Innovation and Societal Transitions* 52 (2024): 100854, doi:https://doi.org/10.1016/j.eist.2024.100854.

4.2. Avogadro, N. and Redondi, R., "Demystifying Electric Aircraft's Role in Aviation Decarbonization: Are First-Generation Electric Aircraft Cost-Effective?" *Transportation Research Part D: Transport and Environment* 130 (2024): 104191, doi:https://doi.org/10.1016/j.trd.2024.104191.

4.3. Timmons, D. and Terwel, R., "Economics of Aviation Fuel Decarbonization: A Preliminary Assessment," *Journal of Cleaner Production* 369 (2022): 133097, doi:https://doi.org/10.1016/j.jclepro.2022.133097.

4.4. Liu, X., Lin, G., Guo, J., Liang, X. et al., "Design and Analysis of a Novel Thermal Management System for Serving Next-Generation Aircraft with Megawatt-Level Heat Loads," *Case Studies in Thermal Engineering* 61 (2024): 104884, doi:https://doi.org/10.1016/j.csite.2024.104884.

4.5. Clima Drift, "The Sustainable Aviation Fuel (SAF) Solution Framework," accessed November 12, 2024, https://www.climatedrift.com/p/the-sustainable-aviation-fuel-saf.

4.6. Escobar, N., Seber, G., Skalsky, R., Wögerer, M. et al., "Spatially-Explicit Land Use Change Emissions and Carbon Payback Times of Biofuels under the Carbon Offsetting and Reduction Scheme for International Aviation (CORSIA)," *Science of the Total Environment* 948 (2024): 174635, doi:https://doi.org/10.1016/j.scitotenv.2024.174635.

4.7. Ecoact, "Navigating CORSIA: An Introduction to Carbon Offsetting and Reduction for International Aviation," accessed November 15, 2024, https://eco-act.com/blog/navigating-corsia/.

4.8. Zhang, J., Zhang, S., Wu, R., Duan, M. et al., "The New CORSIA Baseline Has Limited Motivation to Promote the Green Recovery of Global Aviation," *Environmental Pollution* 289 (2021): 117833, doi:https://doi.org/10.1016/j.envpol.2021.117833.

4.9. Khalifa, R., Alherbawi, M., Elomri, A., and Al-Ansari, T., "Alternative Fuels' Blending Model to Facilitate the Implementation of Carbon Offsetting and Reduction Scheme for International Aviation," *Fuel* 326 (2022): 124974, doi:https://doi.org/10.1016/j.fuel.2022.124974.

4.10. The International Air Transport Association (IATA), "Net Zero 2050 Progress Tracking Methodology," accessed November 15, 2024, https://www.iata.org/contentassets/b3783d24c5834634af59148c718472bb/net-zero-tracking-progress-methodology.pdf.

4.11. The International Civil Aviation Organization (ICAO), "Sustainable Aviation Fuels (SAF)," accessed December 12, 2024, https://www.icao.int/environmental-protection/pages/SAF.aspx.

4.12. JM Johnson Matthey, "Why Sustainable Aviation Fuel (SAF) Is Gaining Altitude: Five Key Drivers of Sustainable Aviation Fuel Production," accessed December 12, 2024, https://matthey.com/media/2023/why-sustainable-aviation-fuel-saf-is-gaining-altitude-five-key-drivers-of-sustainable-aviation-fuel-production.

4.13. GOV.UK, "Sustainable Aviation Fuel (SAF) Mandate," accessed December 13, 2024, https://www.gov.uk/government/collections/sustainable-aviation-fuel-saf-mandate.

4.14. Climate Catalyst, "Sustainable Aviation Fuel Policy in the UK," accessed March 10, 2024, https://climatecatalyst.org/learning-hub/sustainable-aviation-fuel-policy-in-the-uk/.

4.15. European Union, "Regulation (EU) 2023/2405 of the European Parliament and of the Council on Ensuring a Level Playing Field for Sustainable Air Transport (ReFuelEU Aviation)," accessed November 7, 2024, https://eur-lex.europa.eu/eli/reg/2023/2405/oj/eng.

4.16. The International Air Transport Association (IATA), "ReFuelEU Aviation Handbook," accessed October 25, 2024, https://www.iata.org/contentassets/d13875e9ed784f75bac90f000760e998/refuel-eu-aviation-handbook.pdf.

4.17. Winchester, N., Malina, R., Staples, M.D., and Barrett, S.R.H., "The Impact of Advanced Biofuels on Aviation Emissions and Operations in the U.S.," *Energy Economics* 49 (2015): 482-491, doi:https://doi.org/10.1016/j.eneco.2015.03.024.

4.18. Male, J.L., Kintner-Meyer, M.C.W., and Weber, R.S., "The U.S. Energy System and the Production of Sustainable Aviation Fuel from Clean Electricity," *Front. Energy Res.* 9 (2021): 765360, doi:https://doi.org/10.3389/fenrg.2021.765360.

4.19. Martinez-Valencia, L., Peterson, S., Brandt, K., King, A.B. et al., "Impact of Services on the Supply Chain Configuration of Sustainable Aviation Fuel: The Case of CO_2e Emission Reductions in the U.S.," *Journal of Cleaner Production* 404 (2023): 136934, doi:https://doi.org/10.1016/j.jclepro.2023.136934.

4.20. Atems, B., "The Response of the U.S. Aviation Industry to Demand and Supply Shocks in the Oil and Jet Fuel Markets," *Transportation Research Interdisciplinary Perspectives* 11 (2021): 100452, doi:https://doi.org/10.1016/j.trip.2021.100452.

4.21. Lexology, "From Here to There: Challenges in Meeting Sustainable Aviation Fuel Targets," accessed October 28, 2024, https://www.lexology.com/library/detail.aspx?g=26930864-3c4c-4772-8601-8474b0e5c8f2.

4.22. Kurzawska-Pietrowicz, P., "Life Cycle Emission of Selected Sustainable Aviation Fuels – A Review," *Transportation Research Procedia* 75 (2023): 77-85, doi:https://doi.org/10.1016/j.trpro.2023.12.010.

4.23. Lilonfe, S., Davies, B., Abdul-Manan, A.F.N., Dimitriou, I. et al., "A Review of Techno-Economic Analyses and Life Cycle Greenhouse Gas Emissions of Biomass-to-Hydrocarbon 'Drop-In' Fuels," *Sustainable Production and Consumption* 47 (2024): 425-444, doi:https://doi.org/10.1016/j.spc.2024.04.016.

4.24. Bullermann, J., Meyer, N.-C., Krafft, A., and Wirz, F., "Comparison of Fuel Properties of Alternative Drop-In Fuels with Standard Marine Diesel and the Effects of Their Blends," *Fuel* 357, no. Part C (2024): 129937, doi:https://doi.org/10.1016/j.fuel.2023.129937.

4.25. Sheppard, T.J., Specht, D.A., and Barstow, B., "Upper Limit Efficiency Estimates for Electromicrobial Production of Drop-In Jet Fuels," *Bioelectrochemistry* 154 (2023): 108506, doi:https://doi.org/10.1016/j.bioelechem.2023.108506.

4.26. Shell, "Sustainable Aviation Fuel," accessed October 29, 2024, https://www.shell.com/business-customers/aviation/the-future-of-energy/sustainable-aviation-fuel.html.

4.27. AirMed Rescue, "Turn On, Drop-In: Sustainable Aviation Fuel," accessed October 30, 2024, https://www.airmedandrescue.com/latest/long-read/turn-drop-sustainable-aviation-fuel.

4.28. The International Civil Aviation Organization (ICAO), "Sustainable Aviation Fuels Guide," accessed October 30, 2024, https://www.icao.int/environmental-protection/Documents/Sustainable%20Aviation%20Fuels%20Guide_100519.pdf.

4.29. Neste, "ReFuelEU: What Is It and How Will It Impact the Aviation Industry?," accessed November 1, 2024, https://www.neste.com/news-and-insights/aviation/refueleu-explained.

4.30. U.S. Department of Energy (DOE), "Sustainable Aviation Fuel," accessed November 2, 2024, https://afdc.energy.gov/fuels/sustainable-aviation-fuel#:~:text=SAF%20can%20be%20blended%20at,how%20the%20fuel%20is%20produced.

4.31. The National Renewable Energy Laboratory (NREL), "Sustainable Aviation Fuel Blending and Logistics," accessed November 2, 2024, https://www.nrel.gov/docs/fy24osti/90979.pdf?utm_source=NREL+Integrated+Mobility+Sciences+Newsletter&utm_campaign=d9c1809af4-EMAIL_CAMPAIGN_2024_02_13_06_02_COPY_01&utm_medium=email&utm_term=0_-491e760416-289817326.

4.32. Emerald, "Flying into a Climate-Friendly Future with Sustainable Aviation Fuel," accessed November 3, 2024, https://emerald.vc/flying-into-a-climate-friendly-future-with-sustainable-aviation-fuel/.

4.33. Chen, R., Yang, H., Wang, K., and Jiang, C., "Impacts of a Sustainable Aviation Fuel Mandate on Airline Competition—Full-Service Carrier vs. Low-Cost Carrier," *Transportation Research Part B: Methodological* 190 (2024): 103098, doi:https://doi.org/10.1016/j.trb.2024.103098.

4.34. Ahmad, S. and Xu, B., "A Cognitive Mapping Approach to Analyse Stakeholders' Perspectives on Sustainable Aviation Fuels," *Transportation Research Part D: Transport and Environment* 100 (2021): 103076, doi:https://doi.org/10.1016/j.trd.2021.103076.

4.35. Nguyen, K.N. and Vuong, H., "A Case Study on Sustainable Aviation Fuel Adaptation by South East Asian Countries: Opportunities, Reality, and the Current Gaps," *Case Studies in Chemical and Environmental Engineering* 10 (2024): 100988, doi:https://doi.org/10.1016/j.cscee.2024.100988.

4.36. Gan, C., Ma, Q., Bao, S., Wang, X. et al., "Discussion of the Standards System for Sustainable Aviation Fuels: An Aero-Engine Safety Perspective," *Sustainability* 15, no. 24 (2023): 16905, doi:https://doi.org/10.3390/su152416905.

4.37. Ding, S., Ma, Q., Qiu, T., Gan, C. et al., "An Engine-Level Safety Assessment Approach of Sustainable Aviation Fuel Based on a Multi-Fidelity Aerodynamic Model," *Sustainability* 16, no. 9 (2024): 3814, doi:https://doi.org/10.3390/su16093814.

4.38. Zhang, C., Hui, X., Lin, Y., and Sung, C.-J., "Recent Development in Studies of Alternative Jet Fuel Combustion: Progress, Challenges, and Opportunities," *Renewable and Sustainable Energy Reviews* 54 (2016): 120-138, doi:https://doi.org/10.1016/j.rser.2015.09.056.

4.39. Beal, C.M., Cuellar, A.D., and Wagner, T.J., "Sustainability Assessment of Alternative Jet Fuel for the U.S. Department of Defense," *Biomass and Bioenergy* 144 (2021): 105881, doi:https://doi.org/10.1016/j.biombioe.2020.105881.

4.40. Chireshe, F., Petersen, A.M., Ravinath, A., Mnyakeni, L. et al., "Cost-Effective Sustainable Aviation Fuel: Insights from a Techno-Economic and Logistics Analysis," *Renewable and Sustainable Energy Reviews* 210 (2025): 115157, doi:https://doi.org/10.1016/j.rser.2024.115157.

4.41. Cui, Q. and Chen, B., "Cost-Benefit Analysis of Using Sustainable Aviation Fuels in South America," *Journal of Cleaner Production* 435 (2024): 140556, doi:https://doi.org/10.1016/j.jclepro.2024.140556.

4.42. Akter, H.A., Masum, F.H., and Dwivedi, P., "Life Cycle Emissions and Unit Production Cost of Sustainable Aviation Fuel from Logging Residues in Georgia, United States," *Renewable Energy* 228 (2024): 120611, doi:https://doi.org/10.1016/j.renene.2024.120611.

4.43. Rojas-Michaga, M.F., Michailos, S., Cardozo, E., Akram, M. et al., "Sustainable Aviation Fuel (SAF) Production through Power-to-Liquid (PtL): A Combined Techno-Economic and Life Cycle Assessment," *Energy Conversion and Management* 292 (2023): 117427, doi:https://doi.org/10.1016/j.enconman.2023.117427.

4.44. Bullerdiek, N., Neuling, U., and Kaltschmitt, M., "A GHG Reduction Obligation for Sustainable Aviation Fuels (SAF) in the EU and in Germany," *Journal of Air Transport Management* 92 (2021): 102020, doi:https://doi.org/10.1016/j.jairtraman.2021.102020.

4.45. The National Renewable Energy Laboratory (NREL), "Sustainable Aviation Fuel (SAF) State-of-Industry Report: State of SAF Production Process," accessed November 5, 2024, https://www.nrel.gov/docs/fy24osti/87802.pdf.

4.46. U.S. Department of Energy (DOE), "Sustainable Aviation Fuel (SAF) Grants," accessed November 7, 2024, https://afdc.energy.gov/laws/13376.

4.47. The International Council on Clean Transportation (ICCT), "The Cost of Supporting Alternative Jet Fuels in the European Union," accessed November 7, 2024, https://theicct.org/sites/default/files/publications/Alternative_jet_fuels_cost_EU_20190320_1.pdf?ref=ctvc.co.

4.48. Bio4A, "Assessing the Business Case for HEFA-Based Sustainable Aviation Fuels," accessed November 8, 2024, https://www.bio4a.eu/2021/03/26/assessing-the-business-case-for-hefa-based-sustainable-aviation-fuels/.

4.49. Doliente, S.S., Narayan, A., Tapia, J.F.D., Samsatli, N.J. et al., "Bio-Aviation Fuel: A Comprehensive Review and Analysis of the Supply Chain Components," *Front. Energy Res.* 8 (2020): 110, doi:https://doi.org/10.3389/fenrg.2020.00110.

4.50. Kim, Y., Moon, I., Kim, J., and Lee, J., "Renewable Natural Gas Value Chain Based on Cryogenic Carbon Capture, Utilization and Storage, and Power-to-Gas for a Net-Zero CO_2 Economy," *Renewable and Sustainable Energy Reviews* 212 (2025): 115425, doi:https://doi.org/10.1016/j.rser.2025.115425.

4.51. Markets Research Future, "Why Sustainable Aviation Fuel Demand Is Soaring," accessed November 8, 2024, https://www.marketresearchfuture.com/news/why-sustainable-aviation-fuel-demand-is-soaring.

4.52. del Monte, D.M., Cruz, P.L., and Dufour, J., "SAF Production from Cameline Oil Hydrotreatment: A Technoeconomic Assessment of Alternative Process Configurations," *Fuel* 324, no. Part B (2022): 124602, doi:https://doi.org/10.1016/j.fuel.2022.124602.

4.53. Chao, H., Agusdinata, D.B., DeLaurentis, D., and Stechel, E.B., "Carbon Offsetting and Reduction Scheme with Sustainable Aviation Fuel Options: Fleet-Level Carbon Emissions Impacts for U.S. Airlines," *Transportation Research Part D: Transport and Environment* 75 (2019): 42-56, doi:https://doi.org/10.1016/j.trd.2019.08.015.

4.54. ADI Analytic, "Greener Skies: Part I: A Jet Fuel and Sustainable Aviation Fuel Market Outlook," accessed November 9, 2024, https://adi-analytics.com/2022/06/22/greener-skies-part-i-a-jet-fuel-and-sustainable-aviation-fuel-market-outlook/.

4.55. Consultancy.eu, "Global Demand for Sustainable Aviation Fuel (SAF) to Explode by 2030," accessed February 19, 2025, https://www.consultancy.eu/news/11402/global-demand-for-sustainable-aviation-fuel-saf-to-explode-by-2030#:~:text=By%202030%2C%20SAF%20demand%20is,SAF%20mandates%20from%20Asia%20Pacific.

4.56. Simpli Flying, "Sustainable Aviation Fuel Market Map 2023," accessed November 12, 2024, https://green.simpliflying.com/p/sustainable-aviation-fuels-saf-market.

4.57. ADS, "A Guide Sustainable Aviation Fuels," accessed November 12, 2024, https://www.adsgroup.org.uk/knowledge/sustainable-aviation-fuels/.

4.58. The International Air Transport Association (IATA), "Disappointingly Slow Growth in SAF Production," accessed December 12, 2024, https://www.iata.org/en/pressroom/2024-releases/2024-12-10-03/.

4.59. The International Air Transport Association (IATA), "Sustainable Aviation Fuel Output Increases, But Volumes Still Low," accessed December 12, 2024, https://www.iata.org/en/iata-repository/publications/economic-reports/sustainable-aviation-fuel-output-increases-but-volumes-still-low/.

4.60. Ficca, A., Marulo, F., and Sollo, A., "An Open Thinking for a Vision on Sustainable Green Aviation," *Progress in Aerospace Sciences* 141 (2023): 100928, doi:https://doi.org/10.1016/j.paerosci.2023.100928.

4.61. The International Civil Aviation Organization (ICAO), "Global Framework for Aviation Alternative Fuels (GFAAF) News and Activities," accessed December 13, 2024, https://www.google.com/maps/d/viewer?mid=1JsMmhJvm27YLD3CRE31v0PjJ_gN6rVdq&ll=40.052767196000104%2C10.534733820073257&z=2.

4.62. The International Civil Aviation Organization (ICAO), "Global Framework for Aviation Alternative Fuels (GFAAF) Initiatives and Projects," accessed December 13, 2024, https://www.google.com/maps/d/viewer?mid=1JsMmhJvm27YLD3CRE31v0PjJ_gN6rVdq&ll=40.052767196000104%2C10.534733820073257&z=2.

4.63. SAF Investor, "Existing and Planned SAF Projects," accessed December 13, 2024, https://www.safinvestor.com/existing-and-planned-saf-projects/.

4.64. The International Civil Aviation Organization (ICAO), "Global Framework for Aviation Alternative Fuels (GFAAF) Ongoing Deliveries and Batch Delivery," accessed December 13, 2024, https://www.google.com/maps/d/viewer?mid=14ALRPxtGBk_hBMXbjTC9wTZzCZZTn61f&ll=39.66903106068176%2C10.030606206414404&z=2.

4.65. The International Civil Aviation Organization (ICAO), "Airports," accessed December 14, 2024, https://www.icao.int/environmental-protection/SAF/Pages/Airports.aspx.

4.66. U.S. Department of Energy (DOE), "New Report on Pathways to Commercial Liftoff for Sustainable Aviation Fuel," accessed December 14, 2024, https://www.energy.gov/articles/us-department-energy-releases-new-report-pathways-commercial-liftoff-sustainable-aviation.

4.67. The European Union Aviation Safety Agency (EASA), "Sustainable Aviation Fuel Figures and Tables," accessed December 15, 2024, https://www.easa.europa.eu/en/domains/environment/eaer/sustainable-aviation-fuels/figures-and-tables.

4.68. Raman, R., Gunasekar, S., Dávid, L.D., Rahmat, A.F. et al., "Aligning Sustainable Aviation Fuel Research with Sustainable Development Goals: Trends and Thematic Analysis," *Energy Reports* 12 (2024): 2642-2652, doi:https://doi.org/10.1016/j.egyr.2024.08.076.

4.69. Yang, Z., Boehm, R.C., Bell, D.C., and Heyne, J.S., "Maximizing Sustainable Aviation Fuel Usage through Optimization of Distillation Cut Points and Blending," *Fuel* 353 (2023): 129136, doi:https://doi.org/10.1016/j.fuel.2023.129136.

4.70. The Federal Aviation Administration (FAA), "Airworthiness Certification of Aircraft," accessed December 16, 2024, https://www.faa.gov/aircraft/air_cert/aw_cert.

4.71. The International Civil Aviation Organization (ICAO), "CORSIA Eligible Fuels," accessed December 16, 2024, https://www.icao.int/environmental-protection/CORSIA/Pages/CORSIA-Eligible-Fuels.aspx.

4.72. UK Ministry of Defence, "DEF STAN 91-091; Turbine Fuel, Aviation Kerosine Type, Jet A-1," London, 2009, accessed December 18, 2024, https://www.jig.org/documents/defstan-91-091-issue-15/.

4.73. Advancing Standards Transforming Market (ASTM), "Fueling the Future of Aviation," accessed December 17, 2024, https://www.astm.org/news/fueling-future-aviation-ja23.

4.74. Wang, Q.-D., Zeng, P., Yao, Q., Liang, J. et al., "An Experimental and Kinetic Modeling Study on the Ignition Kinetics of a Sustainable Aviation Fuel and Its Blends with a Traditional RP-3 Jet Fuel," *Fuel* 380 (2025): 133191, doi:https://doi.org/10.1016/j.fuel.2024.133191.

4.75. Vincent-Hall, T.D., Bergeron, J.G., Eftim, S.E., Lindahl, A.J. et al., "Health Effects of Occupational Exposure to Jet Fuels Used in the Military: A Systematic Review of the Epidemiologic Literature," *Environment International* 196 (2025): 109278, doi:https://doi.org/10.1016/j.envint.2025.109278.

4.76. Advancing Standards Transforming Market (ASTM), "Standard Practice for Evaluation of New Aviation Turbine Fuels and Fuel Additives," accessed December 18, 2024, https://store.astm.org/d4054-22.html.

4.77. Innovate UK Business Connect, "Understanding the ASTM D4054 Process: A Step-by-Step Guide to Jet Fuel Approval," accessed December 18, 2024, https://iuk-business-connect.org.uk/wp-content/uploads/2022/01/UnderstandingtheASTMD4054process.pdf.

4.78. Joint Inspection Group, "The Fuel Specification for Sustainable Blend Components (SBCs)," accessed December 18, 2024, https://kamino.fra1.cdn.digitaloceanspaces.com/jig/app/uploads/2022/03/TID-No-4-SUSTAINABLE-AVIATION-FUELS-1ST-EDITION-2022_03.pdf.

4.79. Yang, J., Xin, Z., He, Q.(S.), Corscadden, K. et al., "An Overview on Performance Characteristics of Bio-Jet Fuels," *Fuel* 237 (2019): 916-936, doi:https://doi.org/10.1016/j.fuel.2018.10.079.

4.80. SKYNRG, "Sustainable Aviation Fuel Certification and ASTM International: What Is It & Why Does It Matter?" accessed December 19, 2024, https://skynrg.com/sustainable-aviation-fuel-certification-and-astm-international-what-is-it-why-does-it-matter/.

4.81. Walkling, C.J., Zhang, D.D., and Harvey, B.G., "Extended Fuel Properties of Sustainable Aviation Fuel Blends Derived from Linalool and Isoprene," *Fuel* 356 (2024): 129554, doi:https://doi.org/10.1016/j.fuel.2023.129554.

4.82. Pires, R., Silva, T.P., Ribeiro, C., Costa, L. et al., "Carbon Footprint Assessment of Microalgal Biomass Production, Hydrothermal Liquefaction and Refining to Sustainable Aviation Fuel (SAF) in Mainland Portugal," *Algal Research* 84 (2024): 103799, doi:https://doi.org/10.1016/j.algal.2024.103799.

4.83. Xu, J., Tan, H., Wu, J., Han, J. et al., "Optimization Method of Heat Transfer Architecture for Aircraft Fuel Thermal Management Systems," *Chinese Journal of Aeronautics* 38, no. 8 (2025): 103452, doi:https://doi.org/10.1016/j.cja.2025.103452.

4.84. OpenPR, "Aircraft Fuel System Market Share, Trends, Growth Factors, Analysis, Opportunities and Rapid Growth by 2023-2030," accessed December 21, 2024, https://www.openpr.com/news/3029988/aircraft-fuel-system-market-share-trends-growth-factors.

4.85. Islam, S.R., Jiang, L., Swinney, R., Van Horn, D. et al., "Global Combustion Characteristics of a Viscous Alternative Jet fuel and Conventional Jet A-2 Using a Novel Twin-Fluid Injector," *Fuel* 386 (2025): 134303, doi:https://doi.org/10.1016/j.fuel.2025.134303.

4.86. Xu, Z., Wang, M., Chang, L., Pan, K. et al., "Assessing the Particulate Matter Emission Reduction Characteristics of Small Turbofan Engine Fueled with 100% HEFA Sustainable Aviation Fuel," *Science of the Total Environment* 945 (2024): 174128, doi:https://doi.org/10.1016/j.scitotenv.2024.174128.

4.87. AOPA, "How It Works: Aircraft Fuel System," accessed December 21, 2024, https://www.aopa.org/news-and-media/all-news/2020/january/flight-training-magazine/ol-how-it-works-fuel-system.

Chapter 05

Future of SAFs

To achieve sustainable growth, the aviation sector needs to meet today's demands without depleting resources for future generations [5.1]. The aviation sector is aware of its environmental impacts and contribution to climate change and integrates solutions in its future projections. It is actively working on all parts of the value chain on the path to net-zero carbon emissions. To achieve its net-zero carbon emissions target, the aviation sector is prioritizing the reduction in emissions at their source through innovative propulsion technologies and other efficiency improvements, such as enhanced air traffic navigation, while also promoting the use of SAFs [5.2]. The significance of SAFs' contribution toward achieving net-zero carbon by 2050 is illustrated in **Figure 5.1** [5.3].

Figure 5.1 The significance of SAFs' contribution to the net-zero target [5.3].

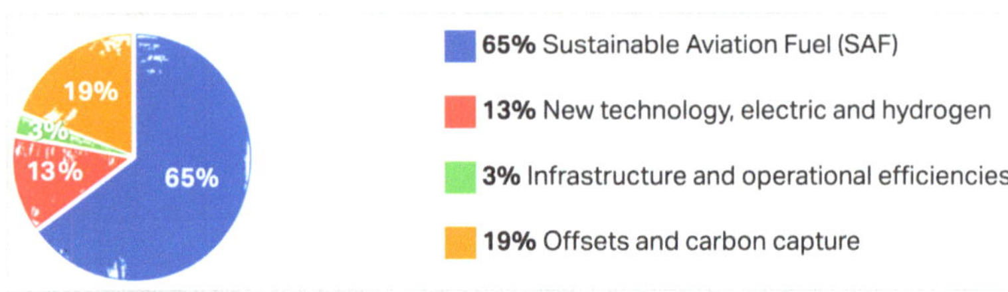

In addition to investing in SAF production (65% SAF) and supporting the development of new technologies for its production (13% new technology), efforts are being made to increase SAF utilization. As the technology and economics of SAFs improve, they hold promise for increasing the scale of use. In recent years, major global aviation corporations have started SAF projects. There is already a commitment by IATA member airlines (covering around 85% of all commercial flight operations) to achieve net-zero carbon emissions from their operations (3% infrastructure) by 2050 [5.3, 5.4]. This commitment aligns (19% carbon capture) air transport with the efforts to support the Paris Agreement's temperature goal. The industry's decarbonization movement has sparked the creation of policies and incentives based on strict minimum GHG emission reduction standards. Through regulations, incentives, and mandates, governments and international organizations are increasingly encouraging the creation and implementation of SAFs. To be successful, coordinated efforts across the entire industry (airlines, airports, air navigation service providers, manufacturers) and significant government support will be required. Despite all regional and economic difficulties, long-term offtake agreements between airlines and SAF suppliers, projects to supply SAFs to all airports, and the use of waste as sustainable raw materials have been implemented [5.5]. These efforts highlight the importance of increasing the use of SAFs in the near future. Thus, various government agencies, national and regional multi-stakeholder initiatives, and individual airlines have set targets for SAF implementation. While there are differences in scope and ambition between individual targets, there is a consensus that SAF use will grow in the coming years. With the policies determined worldwide and the need for global cooperation beyond the aviation sector, an economy has emerged that will enable all countries to participate in the future global SAF market [5.6]. Governments and international aviation organizations have implemented supportive policies and regulations to encourage the use of SAF. Ongoing investments in research and development continue to drive technological improvements, making SAF production more cost-effective. The increasing demand for SAFs by airlines globally presents a significant opportunity for market growth. The future CAGR of SAFs is shown in **Figure 5.2** [5.7]. The global SAF market size has reached USD 1.1 billion according to 2023 data and is expected to reach USD 16.8 billion by 2030, with a CAGR of 47.7% during the forecast period. Increasing concerns over climate change, stringent government regulations to reduce carbon emissions, and reduced dependence on fossil fuels are some of the factors driving the growth of the SAF market. Regulatory initiatives and mandates, particularly from ICAO and various governments, further accelerate market expansion. The economic viability and scale achieved by commercial aviation support the growth of the SAF market size. Increasing the investments in research and development to increase SAF production efficiency, along with advances in feedstock technologies, contributes significantly to the growth and promise of the sector. Collaborations between airlines and biofuel producers play a key role in scaling SAF production and promoting a more sustainable future for air travel.

As shown in **Figure 5.2**, North America will have a large market share in the SAF market in the future. Other countries will adopt a determined and competitive stance in implementing the international policies accepted to achieve a similar economic potential. This competitive advantage will make production even more efficient and cost-effective. The region has a variety of renewable raw materials suitable for SAF production, such as agricultural residues, waste cooking oils, and MSW. This source of raw materials will be widespread and sustainable. North American aviation infrastructure is already well established, and refineries and fuel distribution networks are part of the infrastructure [5.8]. The inclusion of SAFs in the aviation fuel supply chain will be easier to implement. Regarding regional markets, APAC is set to be the second fastest growing region in terms of the SAF market size and CAGR. The economic developments in this region increase the demand for air travel exponentially. The increasing demand for more aviation services will necessitate the development of sustainable fuel options to meet or replace the needs of other fuels. Compared to the North American and European markets, the development of SAFs in the APAC region is still in its infancy and relies heavily on government support [5.9]. This is mainly due to the uneven level of development across countries, which makes it challenging for the region's aviation industry to make the transition. Most of the Asian countries have established policies and regulations with the aim of achieving sustainable development and low carbon emissions. This includes the support for the development and use of SAFs. With respect to the progress in the production capacity and adoption process worldwide as shown in **Figure 5.3**, SAFs have the potential to dominate future aviation applications worldwide with policies supporting them first in North America and then in Asia and Europe [5.10].

Figure 5.2 Global forecast for the SAF market [5.7].

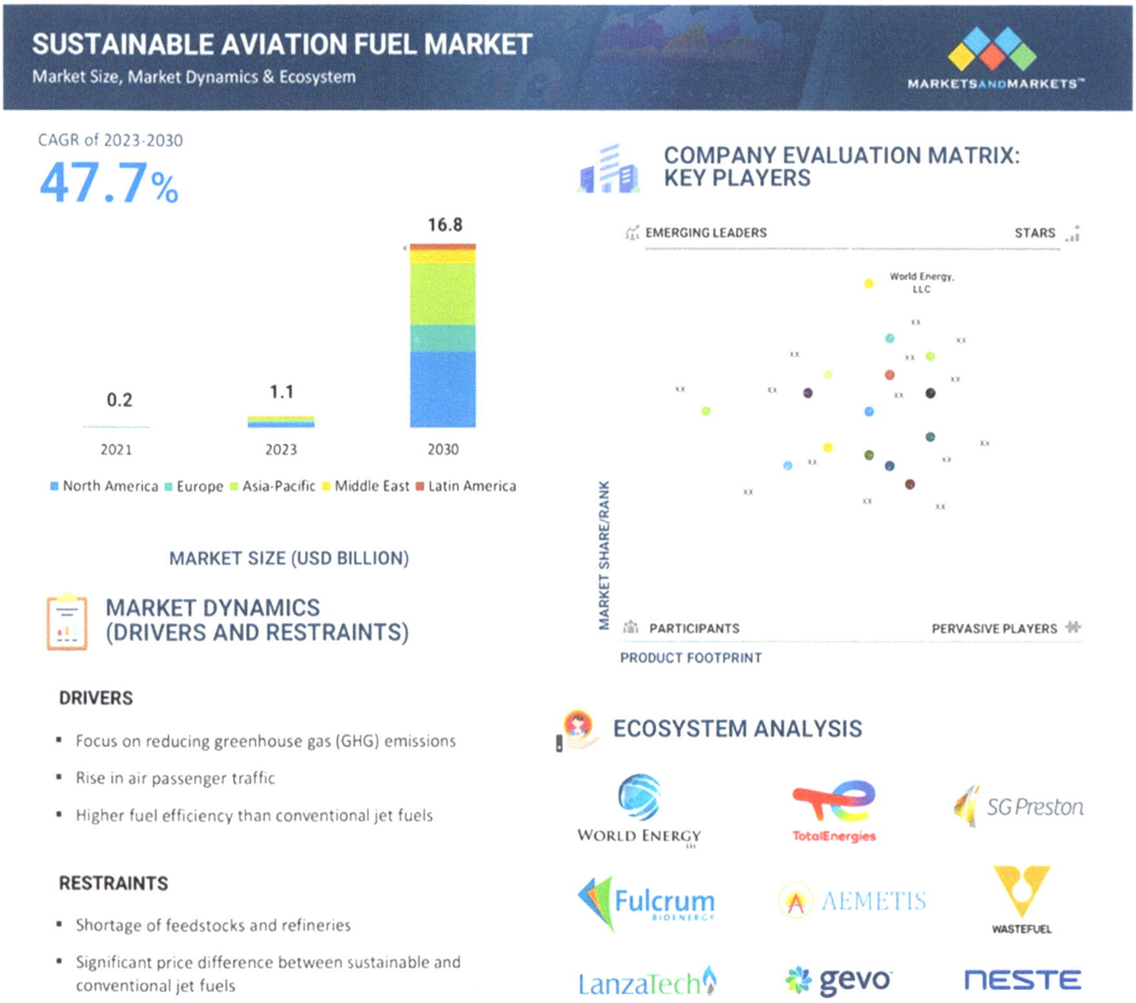

SUSTAINABLE AVIATION FUEL MARKET
Market Size, Market Dynamics & Ecosystem
MARKETSANDMARKETS™

CAGR of 2023-2030
47.7%

16.8
0.2
1.1
2021
2023
2030
North America Europe Asia-Pacific Middle East Latin America

MARKET SIZE (USD BILLION)

MARKET DYNAMICS (DRIVERS AND RESTRAINTS)

DRIVERS

- Focus on reducing greenhouse gas (GHG) emissions
- Rise in air passenger traffic
- Higher fuel efficiency than conventional jet fuels

RESTRAINTS

- Shortage of feedstocks and refineries
- Significant price difference between sustainable and conventional jet fuels

COMPANY EVALUATION MATRIX: KEY PLAYERS

EMERGING LEADERS
STARS
World Energy, LLC
MARKET SHARE/RANK
PARTICIPANTS
PERVASIVE PLAYERS
PRODUCT FOOTPRINT

ECOSYSTEM ANALYSIS

WORLD ENERGY
TotalEnergies
SG Preston
Fulcrum BIOENERGY
AEMETIS
WASTEFUEL

LanzaTech
gevo
NESTE

Source MarketsandMarkets.

Figure 5.3 Worldwide progress in SAF production and adoption [5.10].

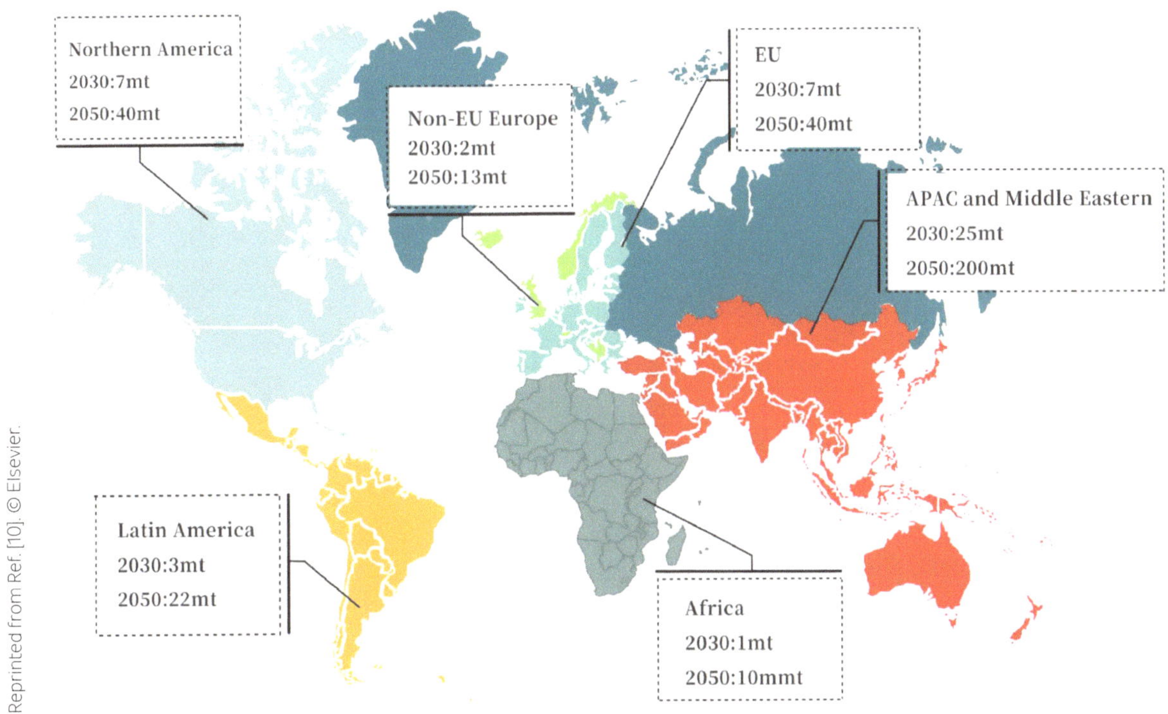

Although the SAF market is growing strongly, in the future efforts will be needed to overcome technical challenges and affordability. Most importantly, the feasibility of SAF production depends on raw feedstock availability, cost-effectiveness, and route selection. Overall, technological advances and gradual cost reductions, along with policy drivers, will be the key factors in encouraging widespread implementation and commercialization of SAFs in the future. As global interest in reducing environmental impacts continues to grow, the demand, production, and market prospects for SAFs are expected to steadily improve. As a result, the SAF market is poised for significant future growth driven by a collective commitment to reducing the aviation industry's carbon footprint and adopting more environmentally friendly alternatives.

Progress and Regulation Status

Decarbonizing air transport is not treated as a single-issue solution for the transport sector in policymaking and execution. It requires a completely new form of problem-solving across governments and industries [5.11]. Decarbonizing air transport is integral to and dependent on the world economy's energy transition and must be addressed as part of national, regional, and global priorities. As shown in **Figure 5.4**, policymaking for biofuels is a multidimensional effort that involves setting clear policy goals across specific stages and timelines [5.12].

Figure 5.4 Stages of policymaking for biofuels [5.12].

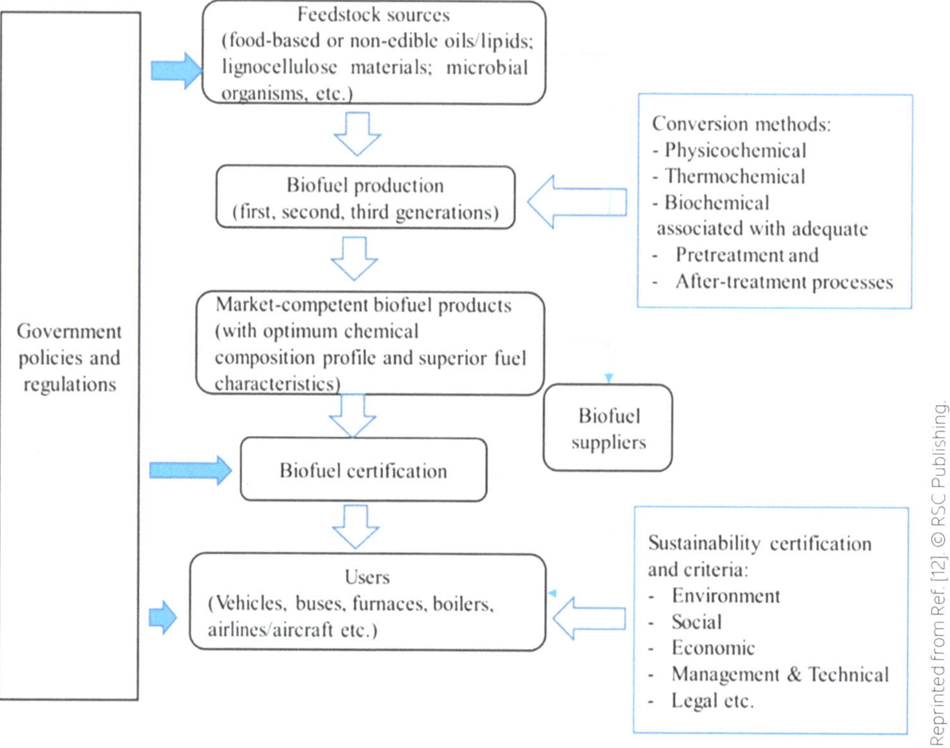

Policy instruments, their mix and intensity, and the order in which they are applied depend on the type of technology, the stage of development, the ability of relevant value chains to scale sufficiently, and country- or region-specific factors [5.13]. Governments at all levels are ensuring that every available lever in the energy transition of air transport is fully utilized and that the necessary tools are in place for the sector to meet its obligations. Therefore, policies are determined to accelerate the creation and development of the necessary new markets and to enable them to fulfill their necessary functions. The policies developed and implemented for using SAFs guide the sustainability of alternative fuels. These policies will help bring together relevant stakeholders and drive investment, market development, and further research to nurture and support a sustainable bioeconomy [5.14]. Effective and sustainable policy incentives are needed to ensure that jet fuel is available to end users at costs comparable to the fossil fuel Jet A and that SAFs are produced profitably. Developing policies that will effectively accelerate the commercial production and distribution of SAFs is a key element in expanding the use of SAFs and managing the process [5.13, 5.14, 5.15]. As air transport becomes a significant part of economies, governments and international organizations are increasingly supporting the development and adoption of SAFs through policies, incentives, and mandates. Most multi-stakeholder initiatives for SAF production and use are well supported by governments [5.16]. Countries considered are those that have already implemented sustainability requirements for biofuels. To date, biofuel regulations have focused on biofuels for road transport rather than aviation use. However, internationally, a number of developed countries are making significant efforts to promote the development of SAFs. The sustainability steps and adaptation process for SAFs are shown in **Figure 5.5** by region and country worldwide [5.17]. Recent policies to support SAF production and increase the demand for SAFs in the US, EU, and internationally are also supported by ICAO and IATA. Commercial aviation is a global industry, and many SAF manufacturers are also global, with operations in the US, EU, and the rest of the world.

Figure 5.5 SAF regulation status of the world [5.17].

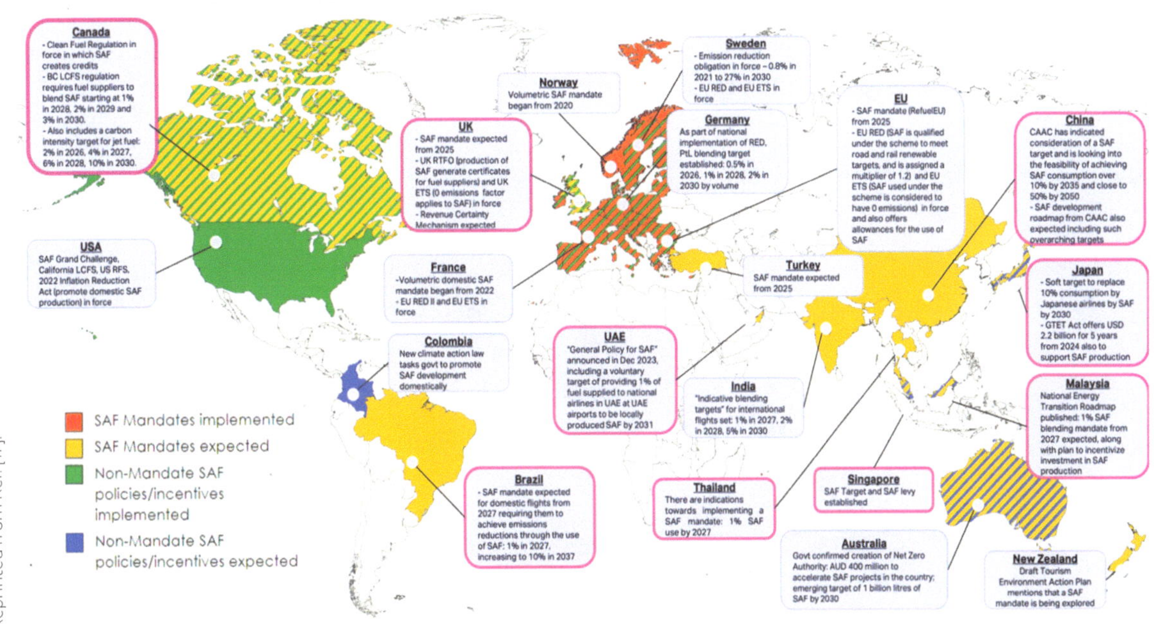

This map reflects regulators in the US, EU, and beyond with plans to scale up SAFs, corporate aviation customers with ambitious plans to purchase SAFs directly, and countries with long-term ambitions [5.17].

The US and the EU, which have played a leading role in the industry, have enacted significant policy legislation supporting the SAF industry. While the objectives of the policies developed in these two regions are similar, their approaches and structures are different. As can be seen in **Figure 5.5**, current legislation in many countries requires compliance with the sustainability criteria for biofuels to be used in air transport, particularly the criteria in the EU RED and the US Renewable Fuel Standard, to benefit from incentives or to be included in the targets set under these legislations. In the US, the federal government implements incentives to reduce SAF costs and increase supply for airlines, while

Europe sets industry targets and fuel blend mandates. The federal government launched a "Sustainable Aviation Fuel Grand Challenge" to reduce cost, increase sustainability, and expand production and use of SAFs [5.18, 5.19]. While current policy and market conditions in the US may encourage biofuel facilities to support production, there is the potential for these same facilities to be used, with some modifications, to increase SAF production and support the federal government's SAF production targets as the SAF market grows. In the Inflation Reduction Act of 2022, the US government proposed two provisions related to SAFs: the SAF blenders tax credit and CFPC [5.20]. Both credits require third-party certification to audit and ensure that producers comply with the policy's sustainability requirements and support programs with ICAO-approved sustainability certification standards. As shown in **Figure 5.6**, the Sustainable Aviation Fuel Grand Challenge projection for the US envisages a 130-fold increase in production scale in the next 7 years and a 12-fold increase in the 20 years after 2030 [5.21]. Producing SAFs to meet the Grand Challenge targets presents a great opportunity for economic development in the US. SAFs have the potential to decarbonize flights within the US and enable US carriers to fly globally with a lower carbon footprint.

Figure 5.6 US Sustainable Aviation Fuel Grand Challenge projection [5.21].

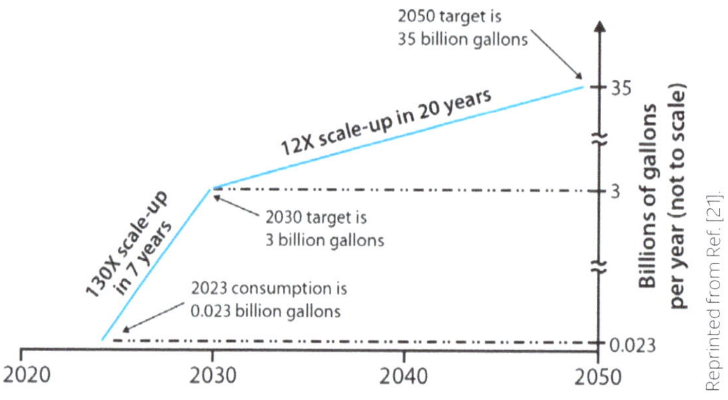

To achieve these ambitious goals, the US Department of Energy, US Department of Agriculture, US EPA, US Department of Transportation, and FAA have collaborated to develop a comprehensive strategy outlined in the Sustainable Aviation Fuel Grand Challenge roadmap. In addition, CAAFI aims to increase energy security and environmental sustainability in aviation through the use of alternative jet fuels. CAAFI is a coalition of airlines, aircraft and engine manufacturers, energy producers, researchers, international participants, and US government agencies. Together, these stakeholders are leading the development and deployment of alternative jet fuels for commercial aviation. The roadmap and other initiatives continue a policy that outlines a "whole of government" approach across all government sectors, with synchronized policies and targeted activities that federal agencies must undertake to achieve the Sustainable Aviation Fuel Grand Challenge's 2030 and 2050 goals [5.22].

State-level policies support SAF production and consumption by providing credits in addition to federal incentives. California stands out as one of the states with the highest economic potential and as a pioneer in initiatives to reduce CI [5.23]. California established the California LCFS in 2009 to develop a range of low-carbon and renewable alternatives to reduce GHG emissions and reduce oil dependency in the transportation sector. As of 2019, SAF is considered an "optional" fuel under the LCFS program. SAF producers can create and sell LCFS credits to "obligated parties" or fossil jet fuel producers to generate revenue. While the LCFS does not mandate a specific blend for jet fuel, the only two major US airports where SAF is regularly used are in California, and the industry consensus is that the LCFS program will continue to encourage SAF production and use in California [5.23, 5.24]. Similar to the California regulations, Washington state's Clean Fuel Standard is designed to encourage a 20% reduction in the CI of transportation fuels below 2017 levels by 2038. Under the bill passed in May 2023, a tax credit incentive for SAF production of USD 1.00 per gallon is provided for SAFs relative to Jet A. Additionally, the incentive package includes a credit for SAF use on flights departing from Washington. States such as Oregon, Minnesota, Washington, and New Mexico have also implemented regulations that provide tax credit for SAF use on state-based flights starting in 2023 and support SAF production [5.25]. Such support benefits state economies and sustainability, enabling progress toward national goals for SAFs. Differences in state policies allow for the development of SAF utilization policies by providing competitive advantages in production capacity and consumption by airlines in different states. Compared to the USA, European airlines have adopted stricter sustainability criteria for biofuels [5.26]. The EU SAF market, incentivized following the adoption of the ReFuelEU aviation regulation and the revision of the EU ETS and the RED, is now in a transition phase. The ReFuelEU aviation initiative is a regulation that is part of the EU's Fit for 55 packages [5.27]. Within the EU, the promotion of renewable transport fuels is mainly regulated by RED II, RED III, and the Fuel Quality Directive (FQD). The European Council formally approved both the ReFuelEU and RED III legislative texts on October 9, 2023. Through RED III, member countries are trying to motivate fuel suppliers to produce more SAF blends through incentives.

The EU points to its ReFuelEU aviation proposal as a policy precursor, a draft regulation that would provide for the full implementation of EU-wide harmonized rules for SAF binding for all member countries. ReFuelEU does provide the final set of rules the industry so desperately needs. The harmonized set of rules across the EU also means that member states cannot impose national obligations to replace ReFuelEU [5.26, 5.27]. Given the international nature of aviation, the EU requires a harmonized set of rules to be applied to all fuel suppliers, airports, and aircraft operators in the EU. The ReFuelEU aviation initiative offers a policy to promote SAFs by gradually increasing the SAF blending requirement and additional requirements for synthetic e-kerosene. According to this regulation, fuel suppliers are required to add SAFs to aviation fuel supplied at EU airports, aircraft operators are required to receive SAF-blended aviation fuel when departing from EU airports, and airport operators are required to provide the necessary infrastructure for the storage and blending of SAFs to fulfill the obligations of fuel suppliers and aircraft operators [5.28]. More than 50% of aviation fuel in Europe is consumed at just ten airports.

The basis of the regulation is that aircraft operators are required by ReFuelEU to meet at least 90% of their annual aviation fuel needs from covered EU airports. This obligation is called an antitanker measure and applies to commercial airlines and all cargo flights. The upgrade obligation for aviation fuel effectively ensures that airlines use up the SAF blends that suppliers are required to deliver. This has led to an approach where fuel suppliers, not member countries or airlines, are obliged to comply with SAF blending targets, even if they supply aviation fuel across country borders. This antitanker measure technically does not prevent aircraft operators from stopping in the UK, Switzerland, or Turkey, where ReFuelEU does not apply, to refuel on long-haul intercontinental routes [5.28, 5.29]. However, stricter regulations have been introduced in policymaking regarding SAF obligations in these countries, so this will no longer be a risk in the future. **Table 5.1** shows the blending ratios of SAFs and synthetic aviation fuels in the future projection in the ReFuelEU proposal of the European Commission and the amendments made in the European Parliament and the Council of the European Union [5.30].

Table 5.1 Projections of SAFs and synthetic aviation fuels by organizations within the EU [5.30].

	European Commission proposal		European Parliament amendments		Council of the European Union amendments	
Year	**Overall SAF target (%)**	**Synthetic subtarget (%)**	**Overall SAF target (%)**	**Synthetic subtarget (%)**	**Overall SAF target (%)**	**Synthetic subtarget (%)**
2025	2	—	2	0.04	2	—
2030	5	0.7	6	2	6	0.7
2035	20	5	20	5	20	5
2040	32	8	37	13	32	8
2045	38	11	54	27	38	11
2050	63	28	85	50	63	28

Fuel suppliers are required to deliver minimum SAF and synthetic aviation fuel blends to the covered EU airports. Starting in 2025, a blend target of 2% SAF is applied, increasing to 70% SAF by 2050. The minimum shares increase in separate steps, while the subtargets have defined averages and minimum shares. This means that fuel suppliers can miss the subtarget and make up for it one year later. As shown in **Figure 5.7**, the EASA conducts a market survey on SAF pricing every year [5.31]. If the minimum usage percentages of SAFs are not met, fuel suppliers are required to pay a fine of at least twice the price difference between conventional aviation fuel and SAF in that year.

Figure 5.7 EASA SAF regulation flow [5.31].

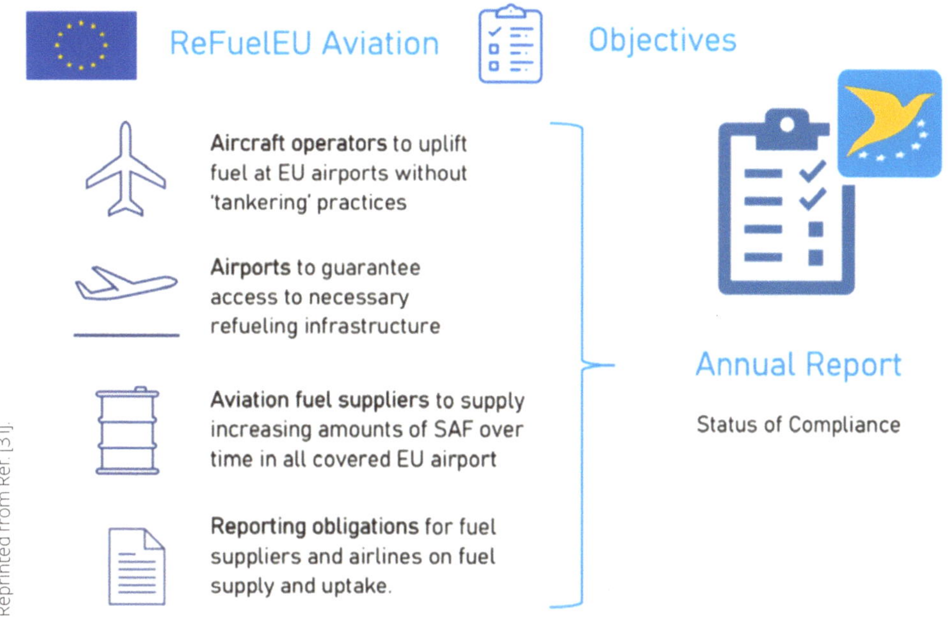

The policy requires a significant expansion of production capacity in order to avoid the EU market becoming overly reliant on imports. Starting in 2025, fuel suppliers are mandated to supply a growing amount of SAFs to EU airports. Similar to the US, the EU aims to introduce a regulation through the Energy Taxation Directive to set minimum tax rates for aviation fuels used on intra-EU flights. The proposed tax rates on aviation fuel will be gradually increased over ten years until a EU-wide minimum rate of 10.75 Euros/GJ is reached, while SAFs are expected to benefit from a zero minimum rate to promote their use [5.31, 5.32]. Another country that regulates the use of SAFs in Europe is the UK. Following Brexit, the UK has not only followed the EU's general policy on aviation carbon reduction but has also published its own Jet Zero Strategy. The UK launched its Transport Carbon Reduction Plan in 2021, setting out its approach to achieving net-zero emissions in aviation by 2050 and net-zero emissions in the UK's domestic aviation by 2040 [5.33]. The proposed SAF usage policy places the responsibility on jet fuel producers rather than airlines. This policy requires them to ensure that at least 10% of their annual production is SAFs by 2030 and to reach a requirement of 75% by 2050. The regulation also requires that fuels produced meet the Ministry of Defence (MOD) Defence Standard (DEF STAN) 91-091 jet fuel specifications [5.33, 5.34]. For aircraft operators, the use of SAFs in the UK is recognized through the UK ETS and ICAO's offset scheme, CORSIA. It is considered that SAF usage targets, which will be implemented in similar time periods across the UK and the EU, will become a common policy for SAF manufacturers and industry across the EU. SAFs are promising solutions for decarbonization in aviation, with global use exceeding 158 million gallons in 2023 and annual production from the SAF sector expecting to reach 500 million gallons by the end of 2025 [5.35]. The US Sustainable Aviation Fuel Grand Challenge effort, Europe's ReFuelEU aviation's ambitious 2% blend target by early 2025, and the UK's ambition to lead SAFs will drive the change in the future. Therefore, the strategic efforts for 2030, followed by net-zero emissions targets by 2050, outlined in **Figure 5.5**, are critical to building a strong international SAF network across the UK, Europe, the Americas, APAC, and MENA. Thus, the future of the SAF industry, its manufacturing development, and its adoption are based on a multifaceted approach that ensures policy stability/longevity and promotes an economically and environmentally sustainable industry. The SAF net-zero emission policy projection determined by IATA is shown in **Figure 5.8** [5.36]. The policy roadmap outlined here highlights the importance of strategic policy sequencing and addresses the need for global collaboration beyond the aviation sector. The main goal of this approach, which includes immediate, medium-, and long-term action plans, is to expand the production and use of SAFs in parallel with ICAO to achieve net-zero emissions. According to this projection, effective SAF policies in the short, medium, and long term should be able to:

- Increase SAF supply.
- Increase the price competitiveness of SAFs.
- Support SAF plant operation.
- Recognize the environmental benefits of SAFs.
- Create structural SAF demand.
- Promote the research and development of new production technology (pathways) and the necessary supply chain.

The success of these policies will play a key role in the aviation sector's transition to net-zero carbon operations. SAF will be an important tool in the aviation sector to achieve net-zero emission targets, with other SAF-oriented policies being developed and implemented worldwide [5.36].

However, efforts from developing countries and the full support of developed countries will be needed to make SAFs economically viable for airlines and to facilitate the increase in commercial capacity that is needed to meet future demands.

Figure 5.8 IATA SAF policy projection for net-zero emission [5.36].

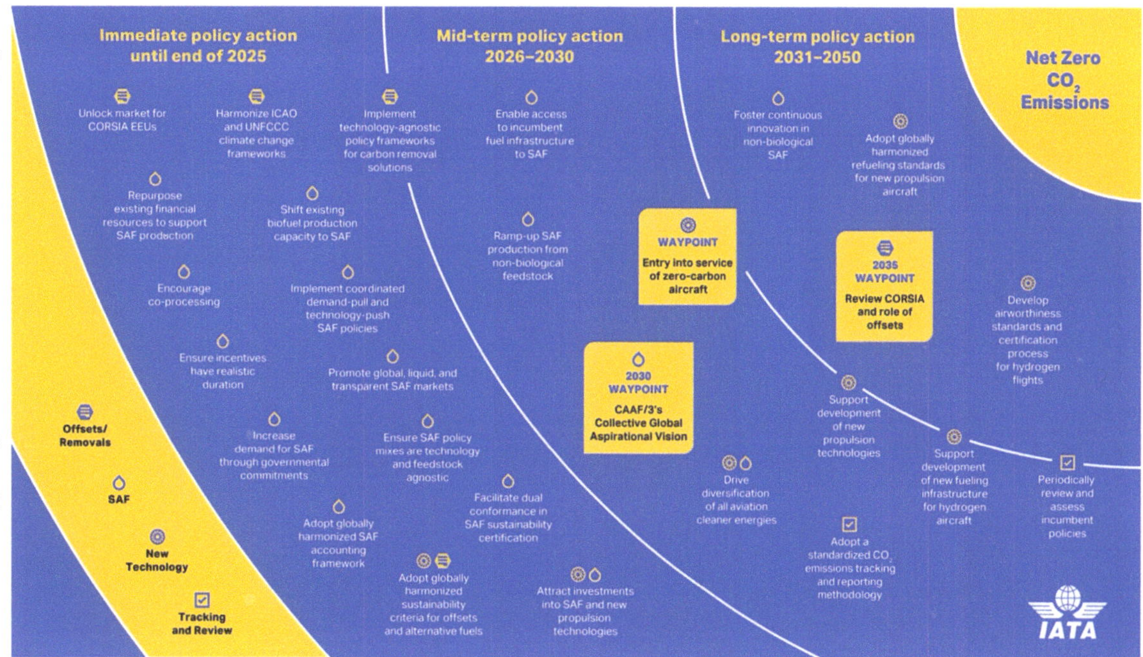

Forthcoming Activities

According to worldwide flight statistics, the total global jet fuel requirement is currently 300+ million tons [5.37]. Taking another important step in the sustainability projection in air transport, the EU has adopted a regulation introducing the Flight Emission Label (FEL) at the end of 2024, which provides a clear and reliable methodology for calculating flight emissions. According to this regulation, airlines flying within the EU or leaving the EU will be able to voluntarily participate in this label, which will be fully operational by July 2025. As shown in **Figure 5.9**, EASA will be responsible for estimating flight emissions in a transparent and harmonized manner, in accordance with the most advanced international standards [5.38, 5.39].

Figure 5.9 FEL regulation for sustainable aviation [5.39].

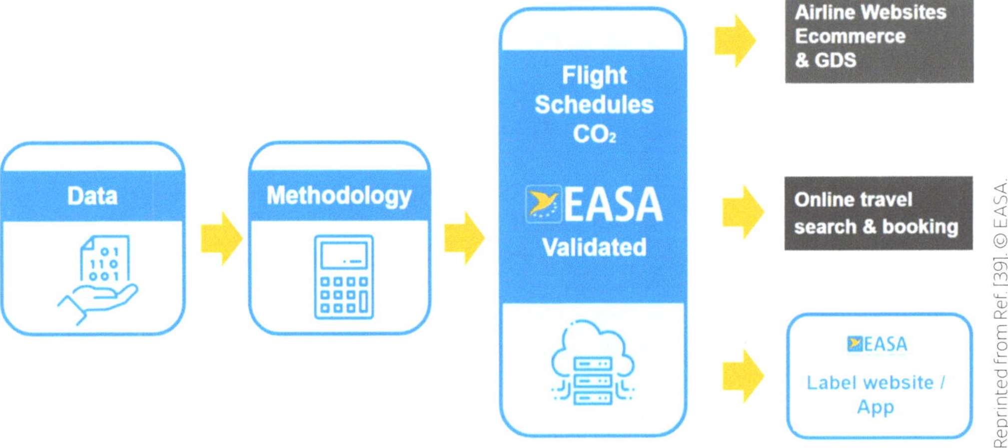

The first flight labels will be assigned by EASA and will be applied to flights planned for the winter season of 2025. The label will provide standardized information on the carbon footprint of flights within the EU. With the increasing need for flights, the FEL will support the transition to net-zero aviation by ensuring fair competition among airlines, encouraging the use of SAFs, and supporting fleet renewal. The FEL is a major initiative that aims to provide passengers with reliable and standardized information on GHG emissions based on actual performance, taking into account factors such as aircraft type, average number of passengers on board, payload volume, and aviation fuel used. Since the emission estimates to be included in this label will be based on real consumption data, it will also encourage airlines to operate more efficiently and sustainably. Due to this and similar important steps to be taken, a 10% target for SAF usage by 2030 would mean that 30+ million tons of demand could be met [5.40]. In terms of future projections for SAFs, SAF-related projects have been ongoing for the past ten years, approximately 190 companies plan to produce the fuel at 330 locations worldwide [5.41]. Of these 190 companies, only 18 are currently actively producing SAFs, and only a few of them are producing on a large scale. Collaborative efforts between countries and organizations will promote standardization and regulatory harmonization to facilitate cross-border acceptance of SAFs. When analyzing these growing stakeholder production demands, traditional energy and chemical companies such as Neste and Shell, with their extensive fuel processing and manufacturing capacities, are among those that provide the foundations for full-scale SAF production. Neste SAF is currently in use at major global airports, including San Francisco International Airport (SFO), Los Angeles International Airport (LAX), Frankfurt Airport (FRA), Amsterdam Airport Schiphol (AMS), Changi Airport (SIN), and Narita International Airport (NRT). These companies will further increase their partnerships in the near future for the use of SAFs in existing airport infrastructure.

The geographical distribution of these current manufacturers is 39% in Europe, 34% in North America, and 27% in the rest of the world. If all existing and candidate producers follow through on their announcements over the next decade and half of their total sustainable fuel production is SAFs, meeting the 10% target by 2030 seems possible [5.41, 5.42].

The SAF dashboard, which is tracked by Boeing as shown in **Figure 5.10**, provides an estimate of worldwide SAF production capacity based on future projections. This provides a projection of worldwide production capacity through 2030.

Figure 5.10 Global SAF cumulative capacity announcements for 2030 [5.43].

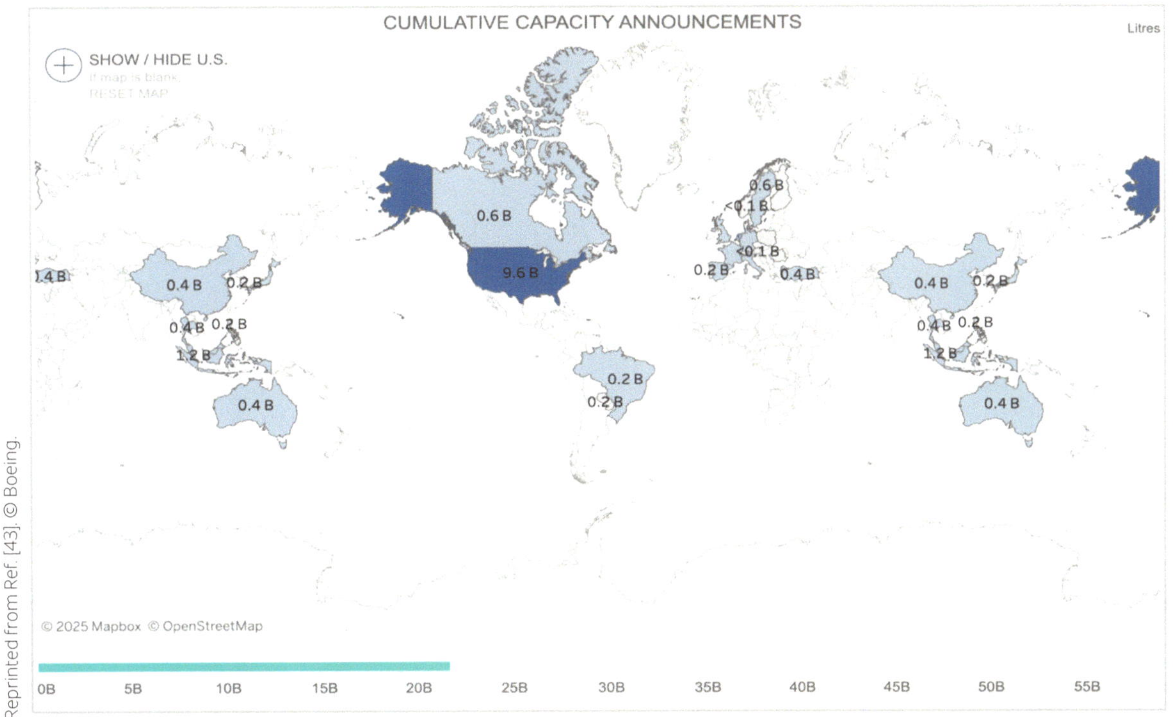

Figure 5.11 shows how much of the projected jet fuel consumption by 2030 would be covered by the declared SAF capacity if SAF were consumed locally by regions around the world.

As shown in **Figures 5.10** and **5.11**, current SAF production capacity of the US and EU is primarily biofuels for the foreseeable future. Additionally, investment and production growth in synthetic jet fuels are included in the 2030 targets to meet synthetic jet fuel subtargets by 2030. Continuous advances in manufacturing technologies, supportive policies, and economic incentives will be important to expand SAF adoption and achieve the sustainability goals of the aviation industry. To achieve this goal and increase production, existing and new SAF production routes need to be scaled up. Scaling up SAF production is a potentially important tool for decarbonizing aviation and offering significant economic and employment opportunities.

Figure 5.11 SAF versus jet fuel consumption for 2030 [5.43].

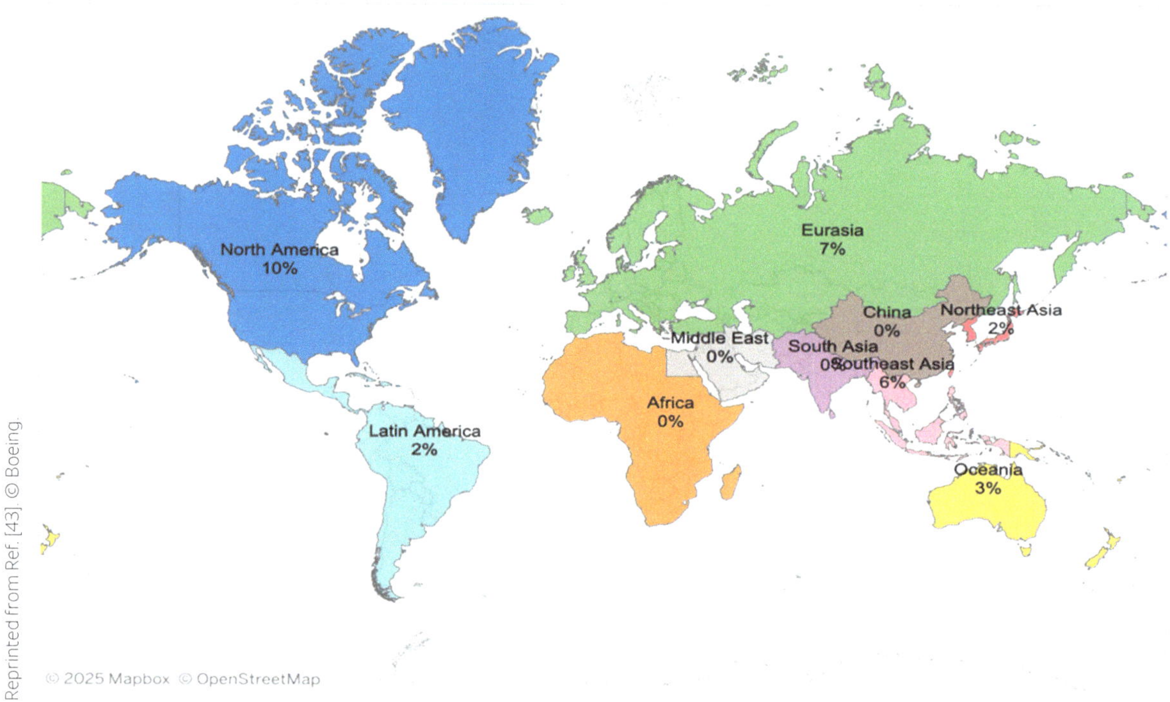

The average total number of jobs associated with the construction and operation of a SAF facility varies depending on the technology path the facility adopts. In general, the more capital-intensive the feedstock supply and production route is, the higher the creation of jobs. Although SAFs have the potential to deliver long-term decarbonization and significant employment benefits, scaling up production to achieve these goals will first require addressing major economic and technological hurdles. Implementing policies that derisk investments in SAF production facilities and implementing public–private partnerships for SAF production and supply will soon become even more important. As a result, the aviation industry using SAFs embodies a transformative force where innovation and responsibility are inter-twined, redefining the future of air travel and taking it to environmentally conscious horizons.

References

5.1. Raab, M., Dietrich, R.-U., Philippi, P., Gibbs, J. et al., "Aviation Fuels of the Future – A Techno-Economic Assessment of Distribution, Fueling and Utilizing Electricity-Based LH2, LCH4 and Kerosene (SAF)," *Energy Conversion and Management: X* 23 (2024): 100611, doi:https://doi.org/10.1016/j.ecmx.2024.100611.

5.2. Dray, L., Schäfer, A.W., Grobler, C. et al., "Cost and Emissions Pathways towards Net-Zero Climate Impacts in Aviation," *Nat. Clim. Chang.* 12 (2022): 956-962, doi:https://doi.org/10.1038/s41558-022-01485-4.

5.3. The International Air Transport Association (IATA), "How We Plan to Achieve Fly Net Zero," accessed January 12, 2025, https://www.iata.org/en/programs/sustainability/flynetzero/.

5.4. Ansell, P.J., "Review of Sustainable Energy Carriers for Aviation: Benefits, Challenges, and Future Viability," *Progress in Aerospace Sciences* 141 (2023): 100919, doi:https://doi.org/10.1016/j.paerosci.2023.100919.

5.5. Dahal, K., Brynolf, S., Xisto, C., Hansson, J. et al., "Techno-Economic Review of Alternative Fuels and Propulsion Systems for the Aviation Sector," *Renewable and Sustainable Energy Reviews* 151 (2021): 111564, doi:https://doi.org/10.1016/j.rser.2021.111564.

5.6. Karanki, F., "Redefining the Skies: How Sustainable Aviation Fuel Alters Airline Cost Structure and Market Dynamics," *International Journal of Sustainable Transportation* 18, no. 8 (2024): 695-703, doi:https://doi.org/10.1080/15568318.2024.2397642.

5.7. Marketa and Market, "Sustainable Aviation Fuel Market Size, Share, Industry Report," accessed January 4, 2025, https://www.marketsandmarkets.com/Market-Reports/sustainable-aviation-fuel-market-70301163.html.

5.8. Mordor Intelligence, "North America Aviation Infrastructure Market Size & Share Analysis - Growth Trends & Forecasts (2025–2030)," accessed January 21, 2025, https://www.mordorintelligence.com/industry-reports/north-america-aviation-infrastructure-market.

5.9. Reuters, "Asia's Green Jet Fuel Ambitions Exceed Demand, Heralding Exports," accessed March 18, 2025, https://www.reuters.com/sustainability/asias-green-jet-fuel-ambitions-exceed-demand-heralding-exports-2025-03-17/.

5.10. Wang, B., Ting, Z.J., and Zhao, M., "Sustainable Aviation Fuels: Key Opportunities and Challenges in Lowering Carbon Emissions for Aviation Industry," *Carbon Capture Science & Technology* 13 (2024): 100263, doi:https://doi.org/10.1016/j.ccst.2024.100263.

5.11. Zheng, S., Wang, C., and Jiang, C., "Carrot or Stick? Environmental and Welfare Implications of Sustainable Aviation Fuel Policies," *Transportation Research Part B: Methodological* 188 (2024): 103062, doi:https://doi.org/10.1016/j.trb.2024.103062.

5.12. Lin, C.-Y. and Lu, C., "Development Perspectives of Promising Lignocellulose Feedstocks for Production of Advanced Generation Biofuels: A Review," *Renewable and Sustainable Energy Reviews* 136 (2021): 110445, doi:https://doi.org/10.1016/j.rser.2020.110445.

5.13. Ebrahimi, S., Esmaeili, S.A.H., Sobhani, A., and Szmerekovsky, J., "Renewable Jet Fuel Supply Chain Network Design: Application of Direct Monetary Incentives," *Applied Energy* 310 (2022): 118569, doi:https://doi.org/10.1016/j.apenergy.2022.118569.

5.14. Proost, S., "Looking for Winning Policies to Address the Climate Issue in EU-Aviation," *Journal of Air Transport Management* 115 (2024): 102534, doi:https://doi.org/10.1016/j.jairtraman.2023.102534.

5.15. Xu, Y., Zhang, Y., Deng, X., Lee, S.-Y. et al., "Bibliometric Analysis and Literature Review on Sustainable Aviation Fuel (SAF): Economic and Management Perspective," *Transport Policy* 162 (2025): 296-312, doi:https://doi.org/10.1016/j.tranpol.2024.11.014.

5.16. Jain, S., Chao, H., Mane, M., Crossley, W.A. et al., "Estimating the Reduction in Future Fleet-Level CO_2 Emissions from Sustainable Aviation Fuel," *Front. Energy Res.* 9 (2021): 771705, doi:https://doi.org/10.3389/fenrg.2021.771705.

5.17. The International Air Transport Association (IATA), "SAF & Net-Zero," accessed February 2, 2025, https://www.icao.int/ESAF/Documents/meetings/2024/Joint%20ICAO%20ESAF%20WACAF%20SAF%20Webinar-%20Programme%20Webinar%2026%20June%202024/ICAO%20ESAF%20WACAF%20SAF%20Workshop%20-%20IATA.pdf.

5.18. U.S. Department of Energy (DOE), "Sustainable Aviation Fuel Grand Challenge October 2021–September 2024 Progress Report," accessed March 3, 2025, https://www.energy.gov/sites/default/files/2025-01/saf-progress-report-2024.pdf.

5.19. Bioenery International, "Sustainable Fuel Federal Agencies Publish SAF Grand Challenge Report," accessed March 3, 2025, https://bioenergyinternational.com/federal-agencies-publish-saf-grand-challenge-report/.

5.20. The U.S. Energy Information Administration (EIA), "U.S. Production Capacity for Sustainable Aviation Fuel to Grow," accessed March 3, 2025, https://www.eia.gov/todayinenergy/detail.php?id=62504.

5.21. The U.S. National Renewable Energy Laboratory, "Sustainable Aviation Fuel (SAF) State-of-Industry Report: State of SAF Production Process," accessed March 5, 2025, https://www.nrel.gov/docs/fy24osti/87802.pdf.

5.22. U.S. Government Accountability Office, "Sustainable Aviation Fuel: Agencies Should Track Progress toward Ambitious Federal Goals GAO-23-105300," accessed March 5, 2025, https://www.gao.gov/products/gao-23-105300.

5.23. Washington State Legislature, "SB 5447.–2023-24: Promoting the Alternative Jet Fuel Industry in Washington," accessed March 6, 2025, https://app.leg.wa.gov/billsummary?BillNumber=5447&Year=2023&Initiative=False.

5.24. U.S. Environmental Protection Agency (EPA), "Renewable Identification Numbers (RINs) under the Renewable Fuel Standard Program," accessed March 6, 2025, https://www.epa.gov/renewable-fuelstandard-program/renewable-identification-numbers-rins-under-renewable-fuel-standard.

5.25. The Office of Governor Michelle Lujan Grisham, "New Mexico Becomes Fourth State to Enact Clean Fuel Standards as Governor Signs Legislation - Landmark Legislation Set to Grow Economy, Reduce Emissions," accessed March 7, 2025, https://www.governor.state.nm.us/2024/03/05/newmexico-becomes-fourth-state-to-enact-clean-fuel-standards-as-governor-signs-legislationlandmark-legislation-set-to-grow-economy-reduce-emissions/.

5.26. Influence Map, "US Sustainable Aviation Fuel (SAF) Policies and Corporate Engagement," accessed March 7, 2025, https://influencemap.org/briefing/-2c7b2a1fb0f08ed81f12681921b0e27f-22997.

5.27. Bardon, P. and Massol, O., "Decarbonizing Aviation with Sustainable Aviation Fuels: Myths and Realities of the Roadmaps to Net Zero by 2050," *Renewable and Sustainable Energy Reviews* 211 (2025): 115279, doi:https://doi.org/10.1016/j.rser.2024.115279.

5.28. Köhler, J., Walz, R., Marscheder-Weidemann, F., and Thedieck, B., "Lead Markets in 2nd Generation Biofuels for Aviation: A Comparison of Germany, Brazil and the USA," *Environmental Innovation and Societal Transitions* 10 (2014): 59-76, doi:https://doi.org/10.1016/j.eist.2013.10.003.

5.29. Scheelhaase, J., Maertens, S., and Grimme, W., "Synthetic Fuels in Aviation – Current Barriers and Potential Political Measures," *Transportation Research Procedia* 43 (2019): 21-30, doi:https://doi.org/10.1016/j.trpro.2019.12.015.

5.30. TOPSOE, "Overview of ReFuelEU Aviation Proposal by European Commission and Amendments Put Forward by Parliament and Council," accessed March 7, 2025, https://www.topsoe.com/sustainable-aviation-fuel/saf-study-appendix3.

5.31. The European Union Aviation Safety Agency (EASA), "EASA Publishes Briefing Note on Sustainable Aviation Fuels (SAF) Prices in the European Union," accessed March 8, 2025, https://www.easa.europa.eu/en/newsroom-and-events/news/easa-publishes-briefing-note-sustainable-aviation-fuels-saf-prices.

5.32. Fageda, X. and Teixidó, J.J., "Pricing Carbon in the Aviation Sector: Evidence from the European Emissions Trading System," *Journal of Environmental Economics and Management* 111 (2022): 102591, doi:https://doi.org/10.1016/j.jeem.2021.102591.

5.33. Chandrasekaran, S., Salah, N.B., and Posada, J.A., "European Union's Biomass Availability for Sustainable Aviation Fuel Production and Potential GHG Emissions Reduction in the Aviation Sector: An Analysis Using GIS Tools for 2030," *Computer Aided Chemical Engineering* 52 (2023): 3055-3060, doi:https://doi.org/10.1016/B978-0-443-15274-0.50487-X.

5.34. Wei, T. and Kallbekken, S., "Carbon Leakage from Aviation under the European Union Fit for 55 Policies," *Transportation Research Part D: Transport and Environment* 132 (2024): 104269, doi:https://doi.org/10.1016/j.trd.2024.104269.

5.35. Morgan, M., Wadud, Z., and Cairns, S., "Can Rail Reduce British Aviation Emissions?" *Transportation Research Part D: Transport and Environment* 138 (2025): 104513, doi:https://doi.org/10.1016/j.trd.2024.104513.

5.36. The International Air Transport Association (IATA), "Policy Net Zero CO_2 Emissions Roadmap," accessed March 9, 2025, https://www.iata.org/en/programs/sustainability/reports/policyroadmap2024/.

5.37. Xiao, F., Dobruszkes, F., Mo, H., and Wang, J., "Revisiting Competition and Complementarity in Multiple Airport Systems: An Analysis of Air Routes and Flights," *Transport Policy* 166 (2025): 50-64, doi:https://doi.org/10.1016/j.tranpol.2025.02.024.

5.38. Carroll, J., Brazil, W., Howard, M., and Denny, E., "Imperfect Emissions Information during Flight Choices and the Role of CO_2 Labelling," *Renewable and Sustainable Energy Reviews* 165 (2022): 112508, doi:https://doi.org/10.1016/j.rser.2022.112508.

5.39. The European Union Aviation Safety Agency (EASA), "EU Flight Emissions Label (FEL)," accessed March 8, 2025, https://www.easa.europa.eu/en/domains/environment/eu-flight-emissions-label-fel.

5.40. Aksoy, H., Domene, M.G., Loganathan, P., Blakey, S. et al., "Case Study on SAF Emissions from Air Travel Considering Emissions Modeling Impact," *Transportation Research Interdisciplinary Perspectives* 29 (2025): 101341, doi:https://doi.org/10.1016/j.trip.2025.101341.

5.41. Sustainable Aviation Futures, "Global SAF Producer and Developer Landscape Report," accessed March 9, 2025, https://www.sustainableaviationfutures.com/saf-spotlight/global-saf-producers.

5.42. Chong, C.T. and Ng, J.-H., "1 - Global Biofuel Policies, Legislations, Initiatives, and Roadmaps," in: Chong, C.T. and Ng, J.-H. (eds.), *Advanced Transport Biofuels*, Woodhead Series in Bioenergy (Sawston, UK: Woodhead Publishing, 2025), 1-68, https://doi.org/10.1016/B978-0-443-15879-7.00001-2.

5.43. CASCADE, "Boeing SAF Dashboard Worldwide View," accessed March 10, 2025, https://cascade.boeing.com/perspectives/boeing-saf-dashboard/.

Index

V

W

About the Author

Nadir Yilmaz, Ph.D., P.E.

Courtesy of Nadir Yilmaz.

Nadir Yilmaz is a professor and the department chair of mechanical engineering at Howard University in Washington, DC, as well as the director of the Howard University Transportation Research Center, a licensed professional engineer (PE), and a consultant. Previously, he was a professor and the associate department chair of mechanical engineering at New Mexico Institute of Mining and Technology. He received BS, MS, and PhD degrees in mechanical engineering from Istanbul Technical University, Bradley University, and New Mexico State University, respectively. He is a fellow of the American Association for the Advancement of Science (AAAS), the American Society for Engineering Education (ASEE), the American Society of Mechanical Engineers (ASME), the National Society of Professional Engineers (NSPE), and the Society of Automotive Engineers (SAE International).

Dr. Yilmaz has been a noted author of over 150 peer-reviewed papers in the fields of alternative fuels, emissions, combustion, rocket propulsion, and automotive engineering, resulting in his recognition on the World's Top 2% Scientists List. He was a National Aeronautics and Space Administration (NASA) Academic Mission Services Faculty Fellow and a U.S. Department of Energy Summer Visiting Faculty Fellow at Los Alamos National Laboratory and Sandia National Laboratories.

Dr. Yilmaz serves as the editor-in-chief for the *SAE International Journal of Fuels and Lubricants* and on the editorial board of numerous other international journals. In addition, he has served on the ASEE Board of Directors, the NSPE Board of Directors, the SAE Foundation Board of Trustees, the NSPE Education Foundation Board of Trustees, and the Board of Directors of Society of Automotive Engineers of Turkey. In a governor-appointed position, he previously served on the New Mexico State Board of Licensure for Professional Engineers and Professional Surveyors.

Dr. Yilmaz has been a recipient of numerous other awards from ASEE, ASME, NSPE, SAE and other organizations including SAE International Leadership Citation Award, SAE Top Contributor, SAE International Excellence in Engineering Education Award, SAE Ralph R. Teetor Educational Award, SAE Faculty Advisor Award, NSPE Engineering Education Excellence Award, New Mexico Engineer of the Year Award, and the 2014 NSPE National Young Engineer of the Year Award, awarded to a PE under the age of 35 in the United States.

Dr. Yilmaz speaks fluent Turkish, English, and Spanish.

www.ingramcontent.com/pod-product-compliance
Lightning Source LLC
Chambersburg PA
CBHW050907210326
41597CB00002B/49

* 9 7 8 1 4 6 8 6 0 7 9 1 8 *